T0298971

Repairing Damaged Wildlands
A Process-Oriented, Landscape-Scale Approach

The unique approach to ecological restoration described in this book will appeal to anyone interested in improving the ecological conditions, biological diversity, or productivity of damaged wildlands. Using sound ecological principles, the author describes how these ecosystems are stabilized and directed toward realistic management objectives using natural recovery processes rather than expensive subsidies. An initial emphasis on repairing water and nutrient cycles, and increasing energy capture, will initiate and direct positive feedback repair systems that drive continuing autogenic recovery. This strategy is most appropriate where landuse goals call for low-input, sustainable vegetation managed for biological diversity, livestock production, timber production, wildlife habitat, watershed management, or ecosystem services. No other book provides such a comprehensive strategy for the ecological restoration of any wildland ecosystem, making this an invaluable resource for professionals working in the fields of ecological restoration, conservation biology and rangeland management.

STEVE WHISENANT is Professor of Rangeland Ecology and Management at Texas A&M University. For over 25 years, his teaching, research and consulting activities have addressed a wide range of ecological restoration and natural resource management issues.

Biological Conservation, Restoration, and Sustainability

The world's biological diversity faces unprecedented threats, and the urgent challenge facing the concerned biologist is to understand ecological processes well enough to maintain their functioning in the face of pressures resulting from human population growth. Perhaps the best hope for the future lies in sustainable development, where both the intrinsic value of the natural world and the importance of the services it provides are acknowledged, and the rate at which resources are utilised does not lead to a long-term decline in biological diversity.

Developing an understanding of the scientific and cultural issues facing those seeking to promote the conservation or restoration of species and habitat provides a major challenge to us all. Those concerned with biological conservation and restoration need skills not only in biological disciplines, but also in the political, social, historical, economic and legal framework within which conservation practice must be developed. Books in this series will therefore, wherever possible, consider the issues from a broad perspective, looking at practical managerial aspects, as well as the applied science of conservation and restoration.

Repairing Damaged Wildlands

A Process-Oriented, Landscape-Scale Approach

STEVEN G. WHISENANT

Department of Rangeland Ecology & Management
Texas A&M University

CAMBRIDGE
UNIVERSITY PRESS

CAMBRIDGE UNIVERSITY PRESS
Cambridge, New York, Melbourne, Madrid, Cape Town, Singapore, São Paulo

Cambridge University Press
The Edinburgh Building, Cambridge CB2 8RU, UK

Published in the United States of America by Cambridge University Press, New York

www.cambridge.org
Information on this title: www.cambridge.org/9780521470018

First published 1999
Fifth printing 2005

A catalogue record for this publication is available from the British Library

Library of Congress Cataloguing in Publication data

Whisenant, Steven G. (Steven Gerald), 1950–
    Repairing damaged wildlands : a process-oriented, landscape-scale
approach / Steven G. Whisenant.
        p.   cm.
    Includes bibliographical references (p.   ) and index.
    ISBN 0 521 47001 3 (hardcover : alk. paper)
    1. Restoration ecology.   2. Conservation biology.   3. Range
management.   I. Title.
    QH541.15.R45W48   1999
    333.7.´153–dc21   98-53259 CIP

ISBN 978-0-521-47001-8 hardback
ISBN 978-0-521-66540-7 paperback

Transferred to digital printing 2008

Contents

Contents

Preface

Ever-increasing demands on wildland ecosystems degrade fundamental resources, reduce species diversity, and lessen the availability of goods and services. Our response to these changes has been constrained by apathy and the very real socioeconomic limitations of wildland ecosystems. Despite great effort and many successes, seriously damaged wildlands are now more abundant than ever. Unrealistic costs, declining benefits, and outright failures have plagued the prevailing paradigms for improving seriously degraded wildlands (or wildlands with low productive potential). The misapplication of otherwise effective agronomic strategies and technologies contributed to these problems.

The outlook supports cautious optimism. Around the world, scientists and practitioners of many disciplines have learned a great deal that is useful for improving damaged wildlands. Although the transfer of this information between countries has improved, communications between disciplines is inadequate. Numerous concepts and practices, with implications for repairing damaged wildlands, are available in the literature of agroforestry, intercropping, soil microbiology, hydrology, ecological engineering, nutrient cycling, mineland reclamation, conservation biology, landscape ecology, and other ecological and applied fields. The abundance of ideas, unique terminology, and differing practices have complicated communication between disciplines at a time when each has much to contribute.

Anyone interested in improving ecological conditions, enhancing biological diversity, or increasing the productivity of damaged wildland ecosystems should find this book useful. I focused on strategies that reduce our dependence on expensive subsidies. Therefore, the approach is to repair processes and initiate natural recovery, rather than replace depleted materials. This deemphasis on replacing materials, like nutrients or basic cations, is not intended to suggest that this approach is not useful in other contexts. Rather, the intent is to describe an approach to developing wildland repair strategies that use autogenic processes, with minimal intervention, to achieve realistic management objectives. This approach is most appropriate where landuse goals call for low-input, sustainable vegetation managed for biological diversity, livestock production, timber production, wildlife habitat, watershed management or certain ecosystem services.

I do not describe all possible repair strategies; the combinations are endless. Rather, it was my intent to set the foundation for a comprehensive approach toward designing situation-specific wildland repair strategies. Although we can make significant improvements in damaged ecosystems, that ability does not justify new or continuing damage. Even significantly improved wildlands are inferior to undamaged wildlands in many ways. The best strategy is to prevent damage before it occurs!

This book takes advantage of the cumulative works of many others. Since the evolution of a new paradigm for repairing wildlands is well underway, an important objective of this book is to facilitate that evolution by integrating relevant theoretical ideas and practical applications, from many disciplines, into a process-oriented conceptual framework. Ecological systems, unlike books, do not have discrete units or chapters. The integration within each chapter increases the cohesiveness of related concepts, reduces fragmentation, and conveys the holistic nature of wildland repair efforts. Still, no single chapter is sufficient to develop effective repair strategies. The comprehensive nature of this book makes it useful as a reference for working professionals, graduate students, or upper-level undergraduate students with a basic knowledge of ecology and soils.

Acknowledgments

I deeply appreciate the support of my family, Denise, Laura, and Justin, who tolerated both my physical and mental absences over the last few years. Through the years, my students at Texas A&M University unknowingly contributed to the development of this book. Their questions, and their answers to my questions, provided the initial impetus for this book and continued to influence its development.

Numerous colleagues read parts of this manuscript and greatly improved it with their insightful comments. James Aronson, Michelle Burk, Chris Call, Jim Dobrowolsky, Sam Fuhlendorf, Durant McArthur, Sue Milton, Keith Owens, Ron Robberecht, and Bruce Roundy each improved one or more chapters. Darrell Ueckert reviewed several chapters and helped me in numerous other ways for three decades. Tom Thurow reviewed the entire book and improved my thoughts on numerous important topics. Two anonymous reviewers provided additional insights on many topics. Conversations with Steve Archer, Tom Boutton, David Briske, David Tongway, Richard Hobbs, Fred Smeins, and Xinyuan (Ben) Wu clarified my thoughts on a variety of ecological topics. Andy Crane's help with field activities was indispensable, because it freed me to concentrate on this book.

Owen Martin of Australian Revegetation Cooperation contributed several photos of their unique equipment. Warren Clary and Tom Thurow also provided photographs. Unless otherwise indicated, all photographs are my own. Bob Whitson, who as Head of the Department of Rangeland Ecology and Management at Texas A&M University, allowed me the freedom to work on this relatively long-term project. I also want to thank Alan Crowden and his staff at Cambridge University Press for their patience and assistance.

Steve Whisenant
College Station, Texas, USA
September, 1998

1

Wildland degradation and repair

Introduction

Wildlands are forests, grasslands, savannas, deserts, wetlands, shrublands, marshlands or other extensively managed areas for which a self-sustaining, and usually perennial, vegetation is the management objective. They often have a relatively low productivity and/or produce goods and services with relatively low market values. However, wildlands, which comprise most of the earth's land area, are very important because they provide food, fiber, recreational amenities, contribute to biological diversity, and control the quality and amount of water for many urban and agricultural uses.

Although initial degradation of wildland ecosystems alters species composition, those areas initially retain control over essential resources (i.e., soil, water, nutrients, and organic materials). Degradation becomes more severe as the area loses control over essential resources (Chapin et al., 1997). Seriously damaged wildlands not only lost control over resources, they lost the capacity for self-repair and are unable prevent additional degradation. Thus, they are less resilient to additional stress or damage and provide fewer environmental services (Myers, 1996). As these degrading processes continue, the area crosses a threshold, beyond which it can no longer recover. This is desertification. Once begun, desertification is a dynamic, self-perpetuating process (Tivy, 1990; Thurow, 1991).

Wildland degradation has two components (socioeconomic and biophysical) that complicate its assessment. The expectations of societies or individual managers for the production of goods and services influence perceptions of wildland degradation. Species composition shifts reduce socioeconomic values without negatively affecting its ability to retain essential resources. Biophysical degradation, the primary focus of this book, occurs when wildland ecosystems lose the ability to retain essential resources. Since biophysical degradation usually has an adverse effect on socioeconomic values, it is included in most assessments of degradation. Some assessments consider the degradation of socioeconomic values, while others do not.

Describing the effects of degradation at regional to global scales is complicated by imprecise information, too little information, and the use of numerous, poorly defined categories of degradation. Thus, global estimates of degradation are rough estimates of variously defined categories. Despite variable definitions, they clearly indicate that serious problems exist on a large scale. For example, almost 17% of the world's vegetated area (20 million km^2) became degraded between 1945 and 1990 (WRI, 1992). Nearly 61% of the world's productive drylands were moderately desertified by 1984 and at least 80% of the rangelands in developing countries were desertified (Mabutt, 1984). Over 12 million km^2 are damaged beyond the repair capacity of individual farmers; 3 million km^2 need extensive engineering work; and 10 000 km^2 are beyond any repair (Mabutt, 1984; Tivy, 1990; Harrison, 1992). Each year an additional 60 000 km^2 are irretrievably lost to degradation (UNEP, 1984). Although damage to wildland ecosystems is defined in many ways and is difficult to quantify with precision, it is clearly a major global problem.

Even the most optimistic estimates of worldwide degradation or desertification indicate the need for ecological repair that far exceeds our capacity to repair damaged wildlands with contemporary approaches. Fortunately, it is possible to initiate natural, plant-driven (autogenic) recovery processes that do not require continuing management subsidies, even on the most degraded sites. Our ability to repair damaged ecosystems is a critical element in the management of the world's environment (Dobson, Bradshaw & Baker, 1997). Wildland economies demand minimal management inputs to initiate autogenic repair processes. Repairing the most severely damaged wildlands may require removal of the physical limitations of the degraded landscape with soil surface modifications that help capture and retain water, soil, nutrients, and seed. While these surface modifications are temporary, they can facilitate establishment of vegetation with the potential to improve conditions. Functionally, repair is completed when predisturbance energy capture rates are restored, nutrient export is minimized, and control of water-use efficiency is realized (Breedlow, Voris & Rogers, 1988). From a practical perspective, certain goods or services are required from these repaired ecosystems.

Repairing damaged wildlands requires realistic objectives that consider the extent of damage, ecological potential, land-use goals, and socioeconomic constraints. Since wildland ecosystems are dynamic and constantly changing, rather than static and predictable, it is unrealistic to set predefined species groups as goals. Instead, redirecting essential ecosystem processes toward preferred trajectories should repair damaged wildlands.

Since the number of potential combinations of objectives, approaches, limitations, and wildland types is almost infinite, step-by-step recommendations are seldom useful. The goal of this book is to describe a framework for repairing damaged wildlands that (1) is process-oriented; (2) seeks to initiate autogenic repair; and (3) considers landscape interactions. The suggested approach begins by assessing the functionality of important primary processes (hydrology, energy capture, and nutrient cycling) and by encouraging positive feedback mechanisms that initiate autogenic repair processes. Positive feedbacks support and reinforce change. That change may either be desirable (improving functionality or conditions) or undesirable (declining functionality or conditions). In contrast, negative feedbacks maintain existing conditions by resisting change. Again, we consider these feedback mechanisms desirable when they resist degradation and maintain functionality. Thus, negative feedbacks that maintain degraded functions and resist improvement are undesirable. Recognizing and appropriately directing these feedback mechanisms will significantly improve our ability to repair damaged wildlands. This is an important goal of this book.

Most contemporary wildland repair programs differ from the approach described in this book in three fundamental ways. First, they emphasize the return of structure (e.g., nutrients and selected plant species) rather than the repair of processes (e.g., hydrology, nutrient cycling, and energy capture). Second, they focus on specific sites without considering the landscape context. Third, they view the 'repair' program as the completion, rather than a beginning of natural repair processes. A focus on returning structural components to functionally damaged ecosystems does not necessarily lead to the development of self-repairing wildland ecosystems.

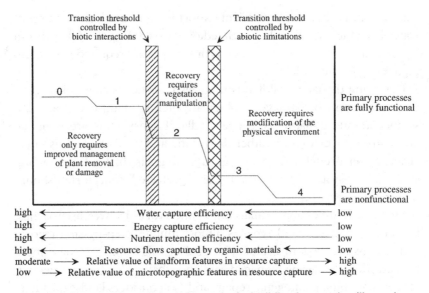

Figure 1.1. Stepwise degradation of hypothetical wildland vegetation illustrating the two common transition thresholds that separate the three vegetative groups emphasized here. Their functional integrity and transition limitations, rather than species composition, define these groups. Wildlands controlled by biotic interactions require some form of vegetation manipulation (some species must be planted while others must be removed) before recovery can occur. Transition thresholds controlled by abiotic limitations require physical manipulations that increase infiltration, reduce erosion, capture organic materials, and/or ameliorate microenvironmental extremes. Vegetative states 0 to 4 follow Milton *et al.* (1994) and are described in Table 1.1.

Degradation

Healthy ecosystems have built-in repair mechanisms, but damage can exceed their capacity for self-repair (Figure 1.1). After crossing that self-repair threshold, natural (unassisted) repair mechanisms cannot repair all the damage. Removing this threshold-related impediment to natural recovery requires active intervention. Our goal is the minimum intervention that removes impediments to autogenic recovery. This does not produce immediate repair; it simply initiates self-repair processes that lead toward properly functioning ecosystems. For our purposes, properly functioning wildlands conserve resources, retain the capacity for self-repair, and provide goods and services that contribute to ecological and socioeconomic sustainability.

Activities that damage and remove vegetation or soil at unsustainable rates damage ecosystem functions. Biomass removal and physical disturbances degrade wildlands. Biomass removal from chronic disturbances (e.g., abusive grazing, fodder removal, or fuelwood collection) damages and kills plants. Acute disturbances remove excessive biomass in single events (e.g., rapid deforestation). Vehicles pack the soil and damage vegetation. Cultivation and mining activities damage and/or remove the soil. Degradation (1) reduces the number of desired plant and animal species; (2) reduces plant biomass; (3) decreases primary production; (4) reduces energy flow to grazing and decomposer components of the food chain; (5) depletes macronutrient pools; and (6) reduces soil stability. Damaged hydrologic, nutrient cycling, energy capture, and vegetation processes contribute to positive feedback systems that increase degradation.

Milton *et al.* (1994) described these changes with a conceptual model of grazing-induced degradation in arid and semiarid ecosystems. They described the symptoms of degradation and suggested focal points for management actions. Thus, it provides a framework for initial damage assessment and preliminary planning of repair strategies (Table 1.1). It is particularly important to recognize the early symptoms of degradation, since management expenses increase with each additional step in the degradation process.

Climatic cycles and stochastic events (Figure 1.2) drive changes on relatively undamaged sites (step 0). Drought, disease, fire, hail, hurricanes, and mudslides cause mass mortality or episodic recruitment that alter species composition and production. Excessive biomass removal over long periods of time usually alters plant populations (Milton *et al.*, 1994), increasing certain species, or life forms, at the expense of others (step 1). The vigor of these frequently defoliated plants is reduced and they produce fewer viable seed. The most effective management option for both these relatively intact areas is adaptive management of the consumers of the ecosystem's primary production. This might involve managing livestock grazing, managing excessive wildlife populations, wood removal, timber harvest, fodder cutting, or other forms of vegetation removal.

With continued overharvest, biological diversity and productivity decrease (step 2) and many of their symbionts and specialized predators are lost (Milton *et al.*, 1994). Reducing plant productivity initiates a series of changes that eventually decrease soil fertility, infiltration rate,

Table 1.1. *Stepwise degradation of wildland landscapes*

Step number	Description	Symptoms	Management options	Appropriate focus for initial repair activities
0	Biomass and composition of vegetation varies with climatic cycles and stochastic events	Perennial vegetation changes are associated with varying climatic conditions (precipitation or temperature) rather than with consumption of primary production. Primary processes are undamaged.	Adaptive management of herbivory, wood harvesting, hay or fodder removal.	Secondary producers (consumers of the ecosystem's primary production) (See Chapter 4)
1	Selective consumption reduces recruitment of most desired plants, allowing populations of less preferred species to expand	Age distribution of plant populations changes to older plants. Primary processes are undamaged.	Stricter control of herbivory, wood harvesting, haying, fodder removal, or other form of selective consumption of plants.	Secondary producers (consumers of the ecosystem's production) (See Chapter 4)
2	Plant species that fail to recruit are lost, as are their specialized predators and symbionts	Plant and animal losses, reduced secondary productivity. Primary processes are damaged, but functioning.	Manage vegetation (e.g., add, remove, or modify) with planting, fire, herbicides, biological, cultural or other methods.	Primary producers (See Chapters 4–7)

3	Biomass and productivity of vegetation fluctuates as ephemerals benefit from loss of perennial cover	Perennial biomass reduced (short-lived plants and instability increase), resident birds decrease, and nomads increase. Primary processes are only partially functional.	Manipulate soil cover (e.g., mulching, erosion barriers, roughen soil surface). Use carefully selected woody vegetation to modify micro-environmental conditions.	Physical environment (See Chapters 2–7)
4	Denudation and desertification involve changes in soil function and detritivore activity	Bare ground, erosion, and aridification. Primary processes are nonfunctional.	Manipulate soil cover (e.g., mulching, erosion barriers, roughen soil surface). Use woody vegetation to modify micro-environmental conditions.	Physical environment (See Chapters 2–7)

Notes:

Symptoms describe the state of plant and animal assemblages, management options refer to actions that a manager could take to repair the site, and the last column refers to the system (level of the food chain) at which management intervention should begin.

Source: Adapted from Milton *et al.*, (1994). Used with permission of the author and the American Institute of Biological Sciences.

Figure 1.2. Relatively intact wildlands – like this in Yellowstone National Park –
retain control over the capture and retention of limiting resources (water, soil,
nutrients, energy, and organic materials). Although step 0 is considered
unchanged and step 1 has undergone changes in species composition and
productivity (Milton *et al.*, 1994), they are functionally similar. Since these areas
are fully functional they retain the capacity for self-repair following disturbance
(such as fire in this example).

and water-holding capacity. Wildlands in this condition (Figure 1.3)
seldom recover naturally without management intervention that adds
and/or removes species. Reversing degradation at this stage has severe
economic restrictions, since it requires both income reductions (fewer
livestock) and expenditures for vegetation manipulation (seeding,
burning, herbicide treatments, or selective plant removal).

Continued reductions in plant productivity decrease litter and vegetative soil cover, which in turn increases erosion and extremes of soil
temperatures (Barrow, 1991). Under these conditions, weedy and
ephemeral species flourish and outcompete seedlings of perennial
plants. Repairing damaged wildlands at this stage (steps 3 and 4) is
unlikely to succeed without addressing the physical limitations of the
degraded landscape. These physical limitations are also important in
the most advanced stage of degradation (Figure 1.4). These sites have
advanced erosion, barren landscapes, are extremely difficult to repair,

Figure 1.3. This Texas site was formerly grassland, but is now dominated by honey mesquite (*Prosopis glandulosa*) and pricklypear (*Opuntia* spp.) and is less productive or produces less of commercial value. Since primary processes are damaged, but still functional, wildlands might be managed to remain at this stage in some situations. This site has passed through a transition threshold that is irreversible without significant management intervention that removes and/or adds plant species.

and recovery may be very slow. Many of these most degraded sites are simply abandoned because repair costs exceed anticipated economic benefits (Barrow, 1991). Fortunately, it is possible to initiate autogenic recovery processes that do not require continuing management subsidies, even on the most degraded sites (Whisenant, Thurow & Maranz, 1995).

Setting realistic objectives

Defining project objectives is the most important single step in the planning process (Pastorok *et al.*, 1997). Specific objectives and knowledge of the economic and biologic restrictions increase the probability of designing and implementing successful repair projects. Repair objectives should specify (1) goals for abiotic functions, performance of

Figure 1.4. This severely degraded landscape in Shaanxi Province, People's Republic of China, is relatively nonfunctional, since it is unable to capture or retain soil, nutrients, water, or organic materials flowing through the landscape. The silted-in reservoir illustrates the magnitude of erosion problems on this landscape. Since little water moves into the soil and there is little vegetation to moderate environmental extremes, it is difficult for plants to become established. Recovery of this site will require physical modifications that reduce abiotic limitations imposed by the lack of vegetation. Despite steep slopes, this area has the ecological potential to develop into forests that stabilize the landscape and retain a high percentage of resource flows. However, socioeconomic pressures (high human population) greatly restrict that option on this landscape.

primary processes, species, communities, and landscape arrangements; (2) landuse, habitat, and/or esthetic goals; (3) spatial scales and time period goals; and (4) performance goals for all important objectives.

Landuse goals, social interactions, economics, management preferences, and biotic and abiotic limitations determine wildland repair objectives. Numerous questions relating to our objectives are important to consider. What are the economic constraints of the program? Must short-term production economics pay for the program or are long-term environmental considerations of overriding importance? Programs designed to restore native vegetation have unique economic environments. Biodiversity programs often emphasize the management and maintenance of ecosystem function and species survival. The unique goals of each program will set the direction of the planning effort.

Damaged wildlands are repaired in many ways and in pursuit of various objectives, but sustainability is the primary objective. The relative importance of social, cultural, economic, or biologic concerns determines our view of sustainability. Sustainable development is 'the maintenance of essential ecological processes and life support systems, the preservation of genetic diversity, and the sustainable use of species and ecosystems' (IUCN, 1983). It is also defined as 'development that meets the needs of the present without compromising the ability of future generations to meet their own needs' (UNEP, 1987). Truly sustainable development requires micro- and macroeconomic evaluations that realistically appraise the environmental consequences of alternative management strategies. Unfortunately, contemporary economic accounting systems seldom consider the adverse environmental impacts of alternative management strategies (Daley, 1991).

What do we call what we want to accomplish?

The literature is complicated by numerous inconsistent definitions of terms that describe the objectives of wildland repair (Table 1.2). These definitions are important because (1) we need clear, well-defined goals; and (2) we should be able to communicate those goals without ambiguity. Unfortunately, we must describe our goals for each situation, since most terms have multiple common uses. Because the literature contains numerous terms, I will briefly review a few of them.

Many restoration efforts seek to return damaged wildlands to some predefined indigenous ecosystem, resembling the original in all respects (Table 1.2). This strict definition of restoration focuses on structure (species), rather than function. This structural focus contributes to ambiguous goals and success criteria (Cairns, 1989; Cairns, 1991). Since we seldom understand the composition, structure, function, or dynamics of historic ecosystems it is difficult to measure success against that goal.

The Society for Ecological Restoration (SER, 1994) went through a relatively rapid change in its concept of ecological restoration (Table 1.2). In three years, their official view of restoration

Table 1.2. *Selected terminology describing objectives for improving ecological condition of damaged wildlands*

Terminology	Definition	Source
Ecological repair	Generic term for improving ecological conditions on damaged wildlands by emphasizing the repair of primary ecosystem processes. Seeks to meet management objectives by developing ecosystems' capacity for self-repair and maintenance.	(NRC, 1974)
Restoration	Return to exact predisturbance conditions.	(SER, 1994)
	The intentional alteration of a site to establish a defined indigenous, historic ecosystem that emulates the structure, functioning, diversity, and dynamics of previous ecosystem (Society for Ecological Restoration in 1990).	(SER, 1994)
	The process of reestablishing to the extent possible the structure, function, and integrity of indigenous ecosystems and the sustaining habitats they provide (Society for Ecological Restoration in 1993).	(SER, 1994)
	The process of repairing damage caused by humans to the diversity and dynamics of indigenous ecosystem (Society for Ecological Restoration in 1994).	(SER, 1994)
	Return of damaged or degraded ecosystem to its presumed or relatively original indigenous state.	(Brown & Lugo, 1994)
	sensu stricto: same as Society for Ecological Restoration in 1990.	(Aronson *et al.*, 1993a)
	sensu lato: 'Endeavors that seek to halt degradation and to redirect a disturbed ecosystem in a trajectory resembling that presumed to have prevailed prior to the onset of disturbance.'	(Aronson *et al.*, 1993a)
Reclamation	Reclaimed site will be similar in ecological functioning and contain similar but not necessarily the same organisms.	(NRC, 1974)
	'Process by which derelict or very degraded lands are returned to productivity and by which some measure of biotic function and productivity is restored.'	(Brown & Lugo, 1994)
Rehabilitation	Making the land useful again but with different land use and different species. Any act of improvement from a degraded state.	(NRC, 1974; Wali, 1992; Bradshaw, 1997) (Aronson *et al.*, 1993a)

	Repair damaged ecosystem functions with the primary goal of raising ecosystem productivity for the benefit of local people. Adopts the indigenous ecosystem's structure and function as the principal model and is directed at recreating self-sustaining system.	(Aronson *et al*., 1993a)
	Return of any damaged or degraded ecosystem to a fully functional ecosystem, without regard to its original or desired final state.	(Brown & Lugo, 1994)
Ecological engineering	'design of human society with its natural environment for the benefit of both.' 'environmental manipulation by man using small amounts of supplemental energy to control systems in which the main energy drives are still coming from natural sources.'	(Mitsch & Jørgensen, 1989) (Odum, 1962)
Reallocation	Landscape is 'assigned new use that does not necessarily bear an intrinsic relationship with the predisturbance ecosystem's structure or functioning. Reallocation assumes a permanent managerial role for people and normally requires ongoing subsidies in the form of energy, water, and fertilizers."	(Aronson *et al*., 1993a)
Reconstruction	Generic term applicable to restoration, rehabilitation or reclamation.	(Allen, 1988a)
Landscape restoration	'The reintroduction and reestablishment of community-like groupings of native species to sites that can reasonably be expected to sustain them, with the resultant vegetation demonstrating aesthetic and dynamic characteristics of the natural communities on which they are based."	(Morrison, 1987)
Landscape rehabilitation	'full or partial placement of structural or functional characteristics that have been extinguished or diminished and the substitution of alternative qualities or characteristics than the ones originally present with the proviso they have more social, economic, or ecological value than existed in the disturbed or displaced state.'	(Cairns, 1988)
Assisted natural restoration	Repair that uses minimal management inputs to stimulate or direct succession.	variously described
Biocultural restoration	Restoration (conservation biology) that explicitly incorporates needs and desires of local inhabitants. 'Explicit and public agreement on management goals is imperative.'	(Janzen, 1988b)

evolved from restoring predefined, indigenous ecosystems (in 1990) to reestablishing the structure, function and integrity of indigenous ecosystems (in 1993) to repairing damage, caused by humans, to the diversity and dynamics of indigenous ecosystems (in 1994). This evolution of terminology reflected contemporary ecological views on succession.

Current ecological theory does not view succession as steady change toward predefined communities in equilibrium with their environment. Rather, it recognizes disturbance-induced discontinuities and irreversible transitions, nonequilibrium communities, and stochastic impacts in succession (Wyant, Maganck & Ham, 1995). In essence, striving to achieve a predefined equilibrium state may be neither possible nor desirable as a management goal (Wyant *et al.*, 1995). Restoration of some predefined ecosystem is unrealistic and/or impossibly expensive (Bradshaw, 1997).

Restoration ecology is a research-oriented discipline that enhances our understanding of ecosystem functioning and provides conceptual direction to manipulative efforts (Table 1.2). Restoration ecology provides a theoretical framework for ecological restoration and makes a valuable contribution by defining ecological principles, testing ecological theories, and facilitating communication between theorists and practitioners.

Rehabilitation (Table 1.2) is usually described as seeking to reduce site degradation and enhance productivity of self-sustaining ecosystems for the benefit of humans (Aronson *et al.*, 1993a). Self-sustaining implies the resilience to recover from any anticipated perturbations, whether human-caused or natural (Aronson *et al*, 1993a). Rehabilitation resembles restoration in that it adopts the indigenous ecosystem's structure and function as much as possible, but without implying perfection (Bradshaw, 1997). It conveys the multiple objectives of halting degrading processes while increasing economic, ecological and esthetic benefits.

Reallocation (Table 1.2) is the conversion to a completely different landuse (Aronson *et al.*, 1993a). This conversion is recommended where the system is seriously degraded and where management objectives or human population pressures necessitate a radically different landuse such as cultivation, improved pasture (irrigated and/or fertilized), agroforestry or other non-wildland uses. Reallocation may require continuing subsidies of fertilizers, herbicides, energy, and water. Reallocation is often essential, but is no longer a wildland system and is not addressed in this book except as it interacts with wildland components within the landscape. These interactions among cultivated fields and wildland components are more fully addressed elsewhere (Aronson, Ovalle & Avendano, 1993c; Hobbs & Saunders, 1993).

Rather than argue over the precise terminology, it seems most useful to emphasize that repair activities occur along a continuum, and that different activities are simply variations of the same theme (Hobbs & Norton, 1996). Repair is a generic term to describe this continuum of objectives (Saunders, Hobbs & Erlich, 1993b; Brown & Lugo, 1994; Whisenant & Tongway, 1995). This book is intended to assist in assessing, planning, implementing, and monitoring these efforts in wildland ecosystems, regardless of specific objectives. Therefore, rather than dwelling on semantics, I will use the term 'repair' because it has broad meaning and suggests a process orientation. My use of the term (repair) implies the goal is the development of a self-repairing ecosystem that meets management objectives by repairing damaged primary processes, and initiating and directing autogenic processes. Placing the emphasis on processes acknowledges the dynamic (rather than static and predictable) nature of ecosystems and the futility of strict species abundance goals (Pickett, Parker & Fiedler, 1992; Pickett & Parker, 1994). This does not mean we repair processes and accept whatever occurs. On the contrary, we apply numerous technologies that direct changes toward management objectives.

Repairing damaged wildlands

Programs designed to improve the ecological status and/or productivity of damaged wildlands usually contain elements of two different approaches (agronomic and ecologic). Although their conceptual approaches differ, both make important contributions toward our understanding of the problems and to the actual repair efforts. The approach described here uses elements of both, but places an emphasis on repairing damaged primary processes and initiating autogenic repair processes on a landscape scale. This approach concentrates on real-world applications that address big problems with few resources by repairing function rather than simply returning structure.

Philosophical approaches

There are numerous philosophical and technological approaches toward improving degraded wildlands. Rather than attempting to describe each of the potential combinations, I will contrast two extremes to illustrate their philosophical differences (Table 1.3). The contrast between agronomic and ecological approaches is somewhat artificial since most repair efforts utilize elements of both, but it illustrates their potential strengths and weaknesses in order to begin discussing synergistic opportunities. The strengths and weaknesses of these approaches are situation specific and neither is universally superior. Successful wildland repair programs generally incorporate some unique combination of both approaches.

AGRONOMIC APPROACH

The philosophical and technological approaches of intensive agricultural endeavors are widely applied to wildland repair efforts, with mixed results. Traditional, agronomic-based approaches toward wildland repair are effective where the soil and climate are most conducive to production. They are also responsible for most of the successful efforts that have occurred in the past. This approach is particularly appropriate at increasing forage production, large-scale projects, and rapid site-stabilization. Modification of traditional farm equipment through several generations produced quality equipment for wildlands. Modified seed drills are now reliable on rocky, unplowed ground and tree transplanters work well on slopes. The quality and variety of equipment available for wildland repair continue to improve.

It is increasingly apparent that when site and environmental conditions are less desirable, the prevailing condition of most wildlands, the benefits of agronomic approaches are often short-lived or not feasible. This situation developed because we attempted to repair wildlands with nutrient subsidies and inorganic and organic amendments rather than by addressing the functioning of the system as a whole. Wildlands are managed as renewable resources with limited subsidies. Sustainable wildland repair strategies must improve the efficiency of resource capture and use within the landscape.

Common shortcomings of the agronomic approach include the possibility of problems due to inefficient nutrient use, poor nutrient retention, narrowed gene pools, low functional diversity, and reliance on elevated management inputs. These agronomic-based strategies are appropriate in some situations, but are impractical on landscapes with marginal productive potential or in developing countries where agricultural chemicals and equipment are unavailable. In relatively predictable conditions, with good edaphic and climatic conditions, this approach can stabilize soils and increase productivity. In less predictable environments, such as arid and semiarid regions, agronomic-based revegetation technologies are less successful because they are neither ecologically based nor economically feasible.

ECOLOGICAL APPROACH

The search for alternative repair strategies and interest in sustainable agriculture stimulated the application of ecological concepts during the repair process. Repair actions initiate a dynamic successional response, toward management goals. Ecologically based approaches direct vegetation change through the enlightened application of ecological principles (Bradshaw, 1983). This approach seeks to create communities and landscapes that persist and develop toward desired conditions. Ecological landscape repair strategies increase and sustain advantageous biological interactions, whereas agronomic approaches typically reduce those biological interactions. Ecological repair strategies do not preclude the use of traditional agronomic practices. The integration of agronomic and ecological practices is very effective.

Ecological strategies modify and enhance soil and microenvironmental conditions with natural processes. The objective is a reduced subsidy approach that uses vegetation suited to existing conditions or vegetation with the ability to improve soil and microenvironmental conditions. Traditional repair efforts often work against normal processes of vegetation change by attempting to maintain artificial communities. Ecologically based approaches often have lower initial investments, but require considerably more time to achieve management goals. Some ecologically oriented projects, particularly in developed countries, are very labor intensive (Cottam, 1987) or equipment intensive and costly (Bruns, 1988). Some programs are implemented with volunteer labor. Governments and private enterprise fund repair programs to mitigate damage caused by mining, construction, or other activities deemed essential to society.

Table 1.3. *Contrasting approaches to the repair of damaged wildlands*

Planning consideration	Agronomic approach	Ecologic approach
Economics	Emphasis on near-term economic return. Suited to conditions where higher probabilities of successful establishment and productive potential allow greater initial investments and continuing management inputs.	Emphasis on long-term ecologic stability and reduced management input. Suited to less favorable or less predictable environments, situations where economic expenditures must be limited, where time is not driving the decision-making process, or where a more stable, natural vegetation complex is the primary objective.
Site selection	Focus on sites with greatest productive potential and greatest probability of returning economic investment.	Focus on sites with greatest potential for meeting ecologic objectives. Those objectives may be soil stabilization, species diversity, structural diversity, functional diversity, or wildlife habitat and do not necessarily include economic performance.
Species selection	Select species that are compatible with management objectives and achieve maximum productivity under the existing soil, climatic, and management environment.	Species may meet objectives either directly or indirectly. Some species may be selected to modify soil or microenvironmental conditions, thus facilitating subsequent recruitment of additional species by natural or artificial means. Species may be used to inhibit the development of other species.
Seedbeds and microsites	Seedbeds are prepared for the site, environmental conditions, and species to be introduced. This may include cultivation, weed control, or chemical and structural changes to the soil.	Species are selected based on adaptation to existing seedbed conditions and may be selected for their site-modifying abilities rather than their ability to accomplish final objectives.

Spatial scale	Repair efforts planned at the stand or individual patch level. This may include single or multiple species mixtures, but seldom includes structural or functional considerations of interacting landscape elements.	Repair efforts planned and implemented at the landscape level to incorporate beneficial structural and functional relationships between interacting landscape elements.
Temporal scale	Benefits begin to be realized relatively rapidly. All species introductions usually occur at the same time.	Initial benefits may be realized less rapidly and full benefits may continue to accrue for years, even decades, as biotic interactions are manifested. Additional species may be introduced in later years.
Ecosystem function	Planning emphasis is to add components of ecosystem structure (such as nutrients or species) to site.	Planning emphasis is to repair ecosystem function.
Maintenance requirements	Moderate	Little or none.

Recommended approach

The approach should be to begin by identifying goals and constraints to those goals. Then we need to assess the status of essential ecological processes and develop alternative strategies to repair each of the identified problems. After assessing the risks of each alternative and their likelihood for success, the complete repair plan is developed (see Chapter 8). Since the potential combinations of unique objectives, approaches, limitations, and wildland types are staggering, step-by-step recommendations for wildland repair are only appropriate for very specific circumstances. A goal of this book is to present a conceptual framework that allows practitioners to develop effective wildland repair programs for any unique combination of circumstances (Figure 1.5). This is most easily accomplished in the wildland context with strategies that (1) are process oriented; (2) seek to initiate autogenic repair; and (3) consider and initiate positive landscape interactions.

PROCESS-ORIENTED STRATEGIES

The recovery and maintenance of processes, rather than species, is the key to ecosystem resilience (Breedlow *et al.*, 1988) and repair (Whisenant, 1995; Whisenant & Tongway, 1995; Bradshaw, 1996). However, wildland repair programs usually emphasize replacing species or nutrients (structure), rather than repairing damaged processes (hydrology, energy capture, nutrient cycling). This book develops a process-oriented approach with an emphasis on managing resource flows and their regulatory mechanisms. This approach begins by assessing the functionality of important primary processes, primarily hydrologic and nutrient cycling (Chapter 2).

Most healthy ecosystems use organic materials to exert and maintain a form of biotic control over nutrient and water flows (Chapin *et al.*, 1997). Degraded ecosystems, with damaged biotic components, have diminished control over these essential hydrologic and nutrient cycling processes. Repairing hydrologic functioning and the mechanisms that regulate resource movement are necessary first considerations during the design of wildland repair strategies. Severely damaged ecosystems have physical limitations to recovery (e.g., steps 3 and 4 in Table 1.1) that are addressed by reducing erosion, protecting the soil surface, increasing infiltration, increasing the water- and nutrient-holding

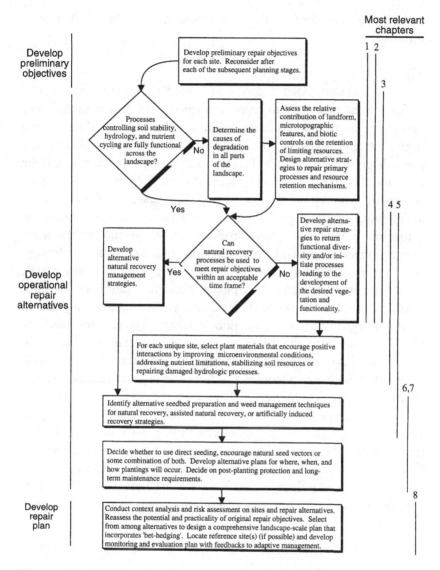

Figure 1.5. Decision tree for planning a process-oriented, landscape-scale wildland repair program. Chapters with useful information at each step are indicated on the right side of the decision tree.

capacity, and ameliorating microenvironmental conditions (Chapters 3 and 4).

INITIATE AND DIRECT AUTOGENIC PROCESSES

Using natural processes to repair damaged wildlands is useful since they are self-sustaining, operate without cost, and are effective on a large scale (Bradshaw, 1996). Since the stability of wildland landscapes depends on stable soils, fully functional hydrologic processes, and the integrity of nutrient cycles and energy flows (NRC, 1994), it is important to focus on repairing those damaged processes. In the long term, maintaining proper functioning of these processes will require the autogenic influences of plant development. Autogenic processes are 'successional change owing to modification of the environment by vegetation (e.g., by producing humus or providing shade)' (Allaby, 1994). Environmental modification by plants is an important component of wildland repair. In contrast, allogenic changes are 'caused by a change in abiotic environmental conditions' (Allaby, 1994). Dysfunctional hydrologic and nutrient cycling processes limit vegetative development that drives autogenic processes (Figure 1.1). Soil surface treatments are needed to change allogenic conditions enough to facilitate subsequent autogenic development (Chapter 3).

Ultimately, vegetation determines the success of wildland repair programs. The selected species must establish, persist, and accomplish management objectives, or facilitate the establishment of additional species that achieve those objectives in the future (Chapter 5). The choice of seedbed preparation (Chapter 6) and planting (Chapter 7) options are especially important because of their long-lasting consequences.

CONSIDER LANDSCAPE INTERACTIONS

Wildland repair programs usually focus on the attributes and objectives of specific pastures, fields, soil types, or ownership units. This limited focus essentially assumes repair sites are functionally isolated from other landscape elements. However, since individual parts of the landscape are continuously gaining and losing water, nutrients, soil, organic materials and propagules, these resource fluxes have important implications. Resource flows, from landscapes with little biomass, are controlled by landform and microtopographic features. For any landform type, the most effective repair strategies manage resource flows by increasing the biotic control over those resources. Since resource fluxes occur in all landscapes, our challenge when planning wildland repair programs is to understand, anticipate, manipulate, and direct resource flows to facilitate the desired changes in ecosystem processes.

Wildland repair programs have the largely unrealized potential to work with underlying landscape processes rather than against them. They accomplish this with strategies that incorporate and direct those processes toward management objectives. Currently, our understanding of landscape function is far from complete. Our ability to direct landscape function is less well developed. However, theoretical, empirical and practical information developed over the last 25 years provides the conceptual foundation for ecological repair of wildland landscapes.

It is imperative to view wildlands from a landscape perspective, but we must not limit our attention to that perspective. Ecological systems are most appropriately viewed as a hierarchy, or a graded series with multiple levels of organization (e.g., organisms, populations, communities, and landscapes) (Archer & Smeins, 1991). Each level in this hierarchy interacts with its physical environment to produce a distinctive functional system with characteristic processes that operate at prescribed spatial and temporal scales. The implication of this ordered perspective is that a complex system may be evaluated without reducing it to a series of simple, disconnected systems. All levels are important and understanding any level requires knowledge of the levels above and below it. The wildland repair approach described in the remaining chapters addresses multiple levels of this hierarchy with assessments and strategies involving individual organisms, populations, communities, and landscapes.

2

Assessing damage to primary processes

Introduction

Wildland ecosystems are biogeochemical systems that use solar energy to convert low-energy inorganic compounds into high-energy organic compounds. Healthy ecosystems use these organic materials to exert and maintain biotic control over flows of soil, water, nutrients, and organic materials (Chapin et al., 1997). Degraded ecosystems, with damaged biotic components, have diminished control over limiting resources (Davenport et al., 1998). In the absence of biotic controls, flows of limiting resources through wildland landscapes are only affected by landform and microtopographic features. A wildland is fully functional when its previous rate of energy capture has been restored, nutrient export has been minimized, and hydrologic processes are properly functioning (Breedlow et al., 1988).

Ecosystems have unique combinations of processes that contribute to proper functioning. Thus, no single assessment is appropriate for all circumstances and management objectives. Criteria that involve the conservation of limiting resources and the functioning of essential primary processes deserve priority status. Since surface soil condition assessments provide insight into stability (ability to withstand erosive forces), hydrologic processes (infiltration and runoff), and nutrient cycling, they provide a useful focal point for repair efforts.

There is ample evidence that replacing species or depleted materials, without repairing damaged primary processes, does not necessarily lead to healthy, self-regulating ecosystems. Our emphasis on function rather than structure requires a focus on the movements of limiting resources (soil, water, nutrients, and organic materials) rather than their abundance. This emphasis on the flows of limiting resources has theoretical support (Finn, 1976) and is a basic tenet of ecological engineering, where external inputs (i.e., forcing functions) are manipulated to direct ecosystem change. Altering forcing functions is believed to cause significant changes in ecosystem structure and function (Jørgensen & Mitsch, 1989). Thus, I suggest an emphasis on repairing damaged energy capture, nutrient cycling, and hydrologic processes, rather than on returning lost materials. This is especially important since ecosystem resilience (ability to recover following disturbance) increases as more energy flows through the system (Loreau, 1994). To achieve this we must distinguish damaged processes from fully functional processes. Then we must assemble a set of assessment attributes that are relevant to the unique problems and objectives of each site.

What is proper versus damaged functioning?

Following minor disturbances in properly functioning ecosystems, biotic recovery mechanisms operate to return and maintain sustainable flows of soil, nutrients, water, and organic materials. With increasing degradation, positive feedback loops develop that both continue and accelerate the impact of damaging processes. Positive feedback systems can destabilize ecosystems, since they reinforce change and cause significant environmental alteration. However, ecosystems with stable equilibria are rare and positive feedbacks are far more important than previously believed (Pahl-Worstl, 1995). Effective wildland repair halts damaging positive feedbacks and initiates positive feedbacks that drive autogenic recovery processes. Since properly functioning wildlands conserve resources, effective wildland repair programs place a strategic emphasis on the processes that efficiently capture and use limiting resources and contribute to proper hydrologic functioning.

Conservation of resources

Landform and microtopographic features control resource flows from severely damaged landscapes with little or no organic materials

(Whisenant & Tongway, 1995). Geomorphic processes and landforms operate at larger scales to control resource flows (Figure 2.1) by creating zones of resource depletion or resource deposition (Toy & Hadley, 1987; Tongway, 1991; Tongway & Ludwig, 1997b). Resource flows usually occur through the fluvial (water) or eolian (wind) transport of soil, water, nutrients, basic cations, and organic materials (Swanson *et al.*, 1988). Similar processes operate at the scale of soil microtopographic depressions (microgilgai, hoofprints, root channels, animal-created holes, cracks in Vertisols, or microcatchments) and aboveground surface obstructions (rocks, logs, vegetation) to conserve resources (Whisenant & Tongway, 1995; Tongway & Ludwig, 1997b). In the absence of effective landform or microtopographic resource control, biotic control mechanisms dominate.

LANDFORM AND GEOMORPHIC PROCESSES

Since processes are difficult to observe, we use landforms to provide visual clues to fluvial and eolian processes. Even a cursory examination of a site's relative topographic position suggests the magnitude of fluvial processes operating across that landscape (Figure 2.1). Relative position within the landform affects runoff rate, water-capture potential of closed basins, and erosion potential. It also suggests the potential for capturing resources from other parts of the landscape. There are at least nine fluvial-created landforms (Dalrymple, Blong & Conacher, 1968), each with unique capacities for capturing or losing resources. Concave sites, of any size, have a relatively high potential to capture resources. Wetlands are an excellent example of concave landforms, and they capture a high percentage of nutrients and organic materials flowing through the landscape. In contrast, without strong biotic controls, convex sites and steep slopes have little control over resource movements.

Where exposed soils (particularly sands) are prevalent, eolian processes dominate to shape the landform and control resource flows. Soil texture provides clues to the relative productive potential of these sites. Dune sands allow deep percolation that protects water from evaporation losses. Thus, desert sands support more vigorous, mesophytic growth than do fine-textured desert soils (Tsoar, 1990). The threshold between the advantage of coarse-textured sand and fine-textured soil lies between 300 mm and 500 mm of precipitation. Below that

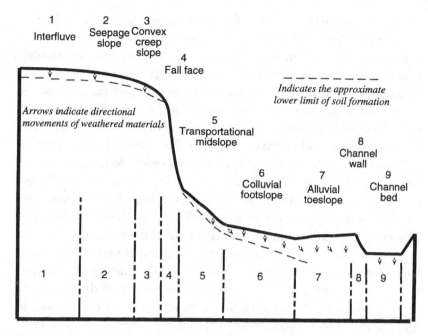

Figure 2.1. Hypothetical arrangement of landforms (from Dalrymple *et al.* 1968) illustrating the relationship of landforms and their relative position with geomorphic processes. Interfluval sites (#1) are dominated by pedogenic processes associated with vertical subsurface soil water movement. Geomorphic processes operating on seepage slopes (#2) are the mechanical and chemical elluviation caused by lateral subsurface water movements. Convex creep slopes (#3) are highly susceptible to soil creep and terracette formation. A fall face (#4) is inevitably dominated by the gravitational transport of most materials and resources. Transporting midslopes (#5) transport material by mass movement (flow, slide, slump, creep), terracette formation, surface and subsurface water action. Colluvial footslopes (#6) are dominated by the redeposition of material through mass movement, surface wash, fan formation, creep and subsurface water action. Alluvial deposition and processes associated with subsurface water movements are most active on alluvial toeslopes (#7). Slumping and falling of materials shapes channel walls (#8). The channel bed (#9) is shaped by the down-valley surface water transport of materials. Used with the permission of Gebrüder Borntraeger (*Zeitschreift Für Geomorphologie*).

threshold, sandy soils are more productive than fine-textured soils (Noy-Meir, 1973).

The shape and arrangement of dunes provide additional insight into their relative stability and repair potential, since actively eroding parts of a dune support little or no vegetation. Dune shape also indicates which part of the dune is most favorable for vegetative establishment. Four major types of dunes occur: transverse, barchan, seif, and vegetated-linear dunes. Transverse and barchan dunes advance by simultaneous erosion on the windward side and deposition on the lee side. Vegetation is found on the crests of transverse and barchan dunes, since sand is neither gained nor lost from the crest. Seif dunes also undergo erosion on one side and deposition on the other. They differ from transverse and barchan dunes in that most seif dune erosion occurs at the crest. Consequently, perennial plants do not grow on the crest or erosional slopes. They are restricted to lower slopes and interdunal areas of seif dunes in humid regions (Tsoar, 1990). Transverse, barchan, and seif dunes are always devoid of vegetation in arid regions (Tsoar, 1990). Only dunes (such as vegetated linear dunes) with less erosion support vegetation in both arid and humid regions. Vegetated linear dunes elongate with little erosion effect. They do not advance, but elongate in proportion to the intensity of dominant winds. Since little erosion occurs on dune crests, they contain perennial plants, even in arid deserts (Tsoar, 1990). Plants capture wind-blown, fine-soil particles that enhance vegetative production by trapping more water and nutrients. These additional resources cause more plant growth, initiating a positive feedback repair system driven by biotic influences.

BIOTIC CONTROLS

As the dominant mechanisms of resource control shift from biotic to abiotic processes, the rate of resource loss increases rapidly (Davenport *et al.*, 1998). Since degraded wildlands typically have less vegetation and organic matter, they have a correspondingly diminished influence over water and nutrient flows within the landscape (Figures 2.2, 2.3). Where organic materials cover much of the soil surface and roots use most of the soil profile, biotic mechanisms are likely to dominate the regulation of resource flows. In healthy grasslands, biologically driven mechanisms regulate resource flows on a very fine scale, with the terrain having less influence (Figure 2.4). As the nature of the system changes, the mechanisms of resource regulation and the scale at which they operate will inevitably change. Where vegetation and topography are not uniform, there is a more variable spatial control over resources.

Changes in the distribution of soil properties may provide an index of relative desertification, at least in arid and semiarid ecosystems

Figure 2.2. Severely degraded site in Niger with crusted soil surface. The movement and retention of water, nutrients and organic materials are primarily regulated by microtopographic and landform features. The few, scattered organic elements of this site exert little biotic control over resource flows. Photograph courtesy of Thomas L. Thurow.

(Tongway & Ludwig, 1994; Schlesinger *et al.*, 1996). The distribution of nutrients, water, and organic matter is patchy in shrublands, with the soil under shrubs serving as resource sinks while the interspaces are sources of limiting resources (Figure 2.5). Following the conversion of desert grasslands to shrublands, the concentration of resources under shrubs has been described as an autogenic process that promotes the persistence of shrubs (Schlesinger *et al.*, 1990). These resource concentrations (also known as fertile islands) are variously viewed as natural triumphs of concentrating biological mechanisms over dispersing physical forces (Garner & Steinberger, 1989), symptoms of degradation (Schlesinger *et al.*, 1996), and tools for repairing severely damaged ecosystems (Whisenant *et al.*, 1995). These different interpretations partially result from a difference of perspective. Conversion of a pristine desert grassland with a relatively uniform distribution of limiting resources (water, nitrogen, and organic matter) to a mesquite (*Prosopis)* sand-dune landscape with a clustered resource distribution involves degradation (Schlesinger *et al.*, 1996). In contrast, severely degraded

Figure 2.3. This severely wind-eroded landscape near Crane, Texas, is nonfunctional. Herbaceous vegetation is gone and 1–2 m of sandy soil has been lost to wind erosion. The few remaining woody plants are unhealthy and capture or retain little of the water, soil, nutrient and organic material flows through the landscape. Water retained on this site is due to the relatively flat terrain and sandy texture of the soil.

ecosystems typically have uniformly low resource levels and high erosion rates. So, conversion from uniformly low resources to a patchy resource distribution following the intentional establishment of woody plants is positive, even if it partially redistributes resources within the landscape (Whisenant *et al.*, 1995).

Soil organic matter is an important biotic regulator of resource flows. Mineral topsoils typically contain only 0.5% to 6.0% organic matter by weight, and subsoils usually contain much less. The soil organic matter content reaches an equilibrium between humus formation (favored by high input rates of residues) and humus loss (favored by moist soils and high temperature) (Loomis & Connor, 1992). Waterlogged soils, with anaerobic conditions, have slow decomposition rates that permit greater organic matter accumulations. Elevated soil temperature leads to rapid decomposition of soil organic matter (Lal & Cummings, 1979). Thus, hot, wet soils tend to have low humus contents while cold, dry climates have greater humus accumulations. If all other factors

Figure 2.4. Largely intact site in the caldenal region of Argentina's pampa. The movement and retention of water, nutrients and organic materials are primarily regulated by biotic components operating on a very fine scale. Landform features exert little control over resource flows.

were the same, tropical soils would have mineralization rates about four times higher than temperate soils (Jenkins & Ayanaba, 1979). Once this equilibrium between the addition and decomposition of organic materials is reached, the amount of humus remains relatively constant unless management actions alter conditions.

Stable soil structure requires continuing organic inputs to the soil (Chepil, 1955). The maintenance, or development, of stable soil aggregates is essential to the proper management or repair of wildland soils. Macroporosity is one of the most important factors in determining how rapidly water moves into soils. Thus, repair strategies that increase aggregate stability play a major role in the return to proper function.

Any disturbance or management regime that reduces vegetative cover and/or raises soil temperature has the potential to significantly reduce soil organic matter. Organic matter levels in soils are reduced by intensive row-cropping, deforestation, abusive grazing management, plowing, accelerated soil erosion, and even industrial waste contamination. These losses of organic matter have long-term consequences. In northeast Colorado, USA, total soil organic carbon (C), nitrogen (N),

Figure 2.5. Shrublands – like this one near Monahans, Texas – have a coarse-scale regulation of limiting resources. The movement and retention of water, nutrients and organic materials are primarily regulated by biotic components operating on a moderately coarse scale (tens of meters). Landform features exert less control over resource flows.

and phosphorus (P) were 55% to 63% lower 60 years after cultivation had stopped (Bowman, Reeder & Lober, 1990). Declines in the labile organic C and N were more rapid and proportionately larger than declines in total organic C and N losses. A high percentage of the short-term (3 years) loss in P occurred from the organic P pool.

Proper hydrologic functioning

Damaged hydrologic functioning drives many of the changes during wildland degradation. Establishing vegetation, conserving resources, repairing nutrient cycling, and increasing energy capture rates require healthy hydrologic functioning. The long-term success of repaired landscapes depends on our ability to assess, recreate, and manipulate hydrologic conditions. Dysfunctional hydrologic processes can cause secondary (human-caused) salinization with serious soil and vegetative implications.

INFILTRATION AND RUNOFF

When water is added (through precipitation, snowmelt or runon) more rapidly than it moves into the soil, the excess water will either pond or runoff. Repairing hydrologic function by increasing infiltration is often the most important step. It often initiates autogenic repair mechanisms leading to recovery of other essential processes. Soil cover, porosity, aggregate stability, preferential flow paths, and biotic (microbiotic) crusts all increase the rate of infiltration (and decrease runoff).

Soil cover protects the soil surface from raindrop impact, reduces surface flows, increases infiltration, and in the long-term improves soil structure (Thurow, 1991). Removing vegetation and litter damages the soil surface and initiates a positive feedback system that accelerates degradation (Figure 2.6). Exposing the soil surface to raindrop impact leads to the development of sealed soil surfaces that greatly reduce the infiltration of water into the soil. A single storm on unprotected soil can reduce infiltration by 50% (Hoogmoed & Stroosnijder, 1984). One study found that crusted soils had infiltration rates 15–20 mm hr^{-1} lower than similar uncrusted soils (Brakensiek & Rawls, 1983). Infiltration rates in Israel were reduced from 100 mm hr^{-1} to 8 mm hr^{-1} on sandy soils and from 45 mm hr^{-1} to 5 mm hr^{-1} on a loess soil (Morin, Benyamini & Michaeli, 1981). Infiltration rates of sandy soils in Mali ranged from 100 mm hr^{-1} to 200 mm hr^{-1} without crusts, but were reduced to 10 mm hr^{-1} after crust formation (Hoogmoed & Stroosnijder, 1984). Reduced infiltration rates of sealed soils significantly decrease the water available for plant growth.

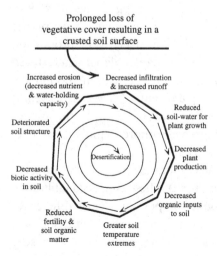

Figure 2.6. Cycle of soil degradation illustrating the importance of the soil surface in the continuing downward spiral of soil condition. Other pathways are possible, but this is most common. While soil surface condition is not the causal factor in all wildland soil degradation processes, it is the most widespread factor. These should be viewed as links in a chain; breaking any link can lead to the initiation of the sequence.

Soil porosity determines the rate at which water moves into the soil, is stored, and made available to plants. Surface soil porosity is one of the first soil attributes to deteriorate during degradation and may be the best functional measure of soil structural condition (Hall, Cannell & Lawton, 1979). The stability of wet soil aggregates measures the resilience of soil structure and influences the maintenance of porosity. Plant species, cultivation, soil type, organic matter, clay content, inorganic ions, climate, and biotic activity influence soil aggregation. Water stable aggregates from 1 mm to 10 mm in diameter are important for plant growth (Tisdale & Oades, 1982), since they produce a mixture of pore sizes that facilitate root growth, water movement, and oxygen diffusion. Soil aggregates should have enough large pores to remain aerobic and enough small pores to hold plant-available water.

Management actions that contribute less organic matter to the soil decrease aggregate stability by exposing the bare soil to raindrop impacts and direct sunlight. This adversely affects infiltration, hydraulic conductivity, erosion potential, and creates less favorable conditions for biological activity (Tisdale & Oades, 1982). Cultivating grassland soils significantly reduces the amount and size distribution of water-stable aggregates (Tisdale & Oades, 1982; Jastrow, 1987). Cultivated soils may need 30–50 years for the size distribution of water-stable aggregates to approach that of uncultivated grassland soils (Tisdale & Oades, 1982). In southern Oregon, 15- to 20-year old clear-cuts that were not reforested still had fewer large soil aggregates than nearby forested sites (Borchers & Perry, 1987). In Illinois, recovery of the larger ($>$0.2 mm and $<$20 mm) water-stable aggregates after long-term cultivation was found to be more closely associated with prairie graminoids than with other vegetation (Jastrow, 1987). The proportion of aggregates greater than 0.2 mm was significantly higher in a restored prairie compared with a nearby ungrazed pasture even though the restored prairie had been cultivated more recently (11 versus 14 years) (Jastrow, 1987). This suggested warm season (C_4) prairie grasses restored the larger water-stable aggregates more rapidly than cool season (C_3) pasture grasses (Jastrow, 1987).

Because soils are rarely uniform, water infiltrates more rapidly into some surfaces than others (Rice & Bowman, 1988). These enhanced infiltration routes, preferential flow paths, produce important spatial differences (Edwards, 1991). Shrinkage cracks in dry montmorillonitic soils are obvious preferential flow paths during high-intensity storms. In soils containing woody plants, preferential flows move into channels around stems and decaying roots (De Vries & Chow, 1978). Macrochannels produced by the aggregation of soil particles into distinct structural units (peds) result in rapid, deep wetting by a slower frontal advance into the peds. Macropores and spaces created by earthworms, previous root growth, burrowing animals, and shrinking clays are the major routes for root, water, and air movements. These macropores drain rapidly, retaining only films of water adsorbed to soil particles by weak hydrogen bonds. The smaller capillary pores are most important as holding spaces for water. Pores <30 µm diameter fall in the capillary range and can hold water against the force of gravity (Loomis & Connor, 1992).

Microbiotic crusts may contain algae, lichens, liverworts, and mosses and be the dominant ground cover in the absence of vascular plants. Algal filaments, fungal mycelia, and tissues of lichens and mosses are often present in the surface few millimeters of these microbiotic crusts. They play an important role in soil stability (Anderson, Harper & Holmgren, 1982; Williams, Dobrowolski & West, 1997), nitrogen fixation (Evans & Ehleringer, 1993), biomass production (Isichei, 1990) and the incorporation of eolian dust into stable soil crusts (Gillette & Dobrowolski, 1993). Soil surface stability is most improved where lichens and mosses are abundant, but filamentous algae are probably more effective in binding the surface soil particles. The thick gelatinous sheaths covering some of the algal species add strength to the web of filaments among the soil particles in the surface 1 mm or 2 mm of a well-established algal crust.

In eastern Australia, soil surfaces with a high cover of microbiotic crusts were more stable and less erodible than surfaces with less microbiotic cover (Eldridge, 1993b). Other studies found microbiotic crusts increased infiltration and decreased sediment loss (Loope & Gifford, 1972). As soil surface conditions deteriorate, microbiotic cover becomes increasingly more important as a descriptor of soil hydrologic properties (Eldridge, 1993b; Eldridge, 1993a). While the benefits of microbiotic crusts to arid soils are clear, evidence from other environments suggests the hydrologic benefits may only occur on degraded soils. In well-structured soils of wooded, semiarid, Australian rangelands, microbiotic cover had no effect on hydrologic conditions (Eldridge, Tozer & Slangen, 1997). Although microbiotic crusts are common in arid, high-elevation deserts of the western United States, they do not occur in the hottest deserts.

EVAPORATION, TRANSPIRATION, AND SALINIZATION
Disruption of a landscape's hydrologic processes triggers unanticipated interactions with catastrophic consequences on other parts of the
landscape. Management actions that result in long-term vegetation
reductions, such as chronic overgrazing or deforestation, seriously
damage hydrologic processes such as evaporation and transpiration.
These vegetation conversions reduce transpiration. These changes can
elevate the water table because shallow-rooted species remove less
water, transpire less water, and intercept less precipitation than the previous vegetation (Greenwood, 1988). Deep-rooted shrubs or trees on
parts of a landscape regulate the hydrology and nutrient retention
capacity for the larger landscape (Ryszkowski, 1989; Ryszkowski, 1992;
Burel, Baudry & Lefeuvre, 1993; Hobbs, 1993).

Understanding how vegetation controls hydrologic processes suggests powerful tools for repairing damaged landscapes. Structural
attributes of vegetation influence interception, transpiration, and water
yields from wildlands (Brooks *et al.*, 1991; Satterlund & Adams, 1992).
Since much of the annual precipitation input, in many wildland ecosystems, is lost through evaporative processes, actions that change vegetation structure can profoundly influence water budgets. For example,
water yields from watersheds typically increase when (1) trees are
removed or thinned; (2) vegetation is converted from deep-rooted
species to shallow-rooted species; or (3) plant cover changes from
species with high interception capacities to species with lower interception capacities (Brooks *et al.*, 1991). The absolute extent of water yield
change is smaller in semiarid regions, but the economic and ecological
importance of that change is often greater.

Wetlands slow the flow of water, increase organic matter accumulation, and have high evapotranspiration rates. Wetland degradation
accelerates the flow of water and increases organic matter decomposition rates. In western White Russia during the 1960s, large wetlands
were drained and converted to cultivation. These hydrologic changes
accelerated organic matter decomposition, increased plant pest problems, and significantly reduced crop yields (Susheya & Parfenov,
1982). These unanticipated problems led to the implementation of new
programs to recreate numerous bogs within those large, cultivated
areas. The bogs helped maintain ecosystem function and minimized
damage done by the large-scale drainage operations.

External influences occasionally disrupt the functioning of internal ecological processes. In these situations, site-specific repair activities are ineffective unless planned to address the landscape-scale problems. For example, salinization is a common result of altered hydrologic processes in arid and semiarid regions and is a major cause of desertification (Grainger, 1992; Thomas & Middleton, 1993). Salinization has several causes: (1) natural drainage and evaporation processes; (2) irrigation in high-evaporation environments; (3) damaged landscape-scale hydrologic processes following major vegetation changes; (4) seawater incursion; or (5) disposal of saline wastes. Those caused by human activities are categorized as secondary salinization. This concentration of salt in the soil surface effectively seals the surface by dispersing soil particles, which reduces pore space, and significantly reduces infiltration. Salinization also reduces the growth rate of plants since energy that would otherwise contribute to growth must be used to extract water.

From an agronomic viewpoint, natural salinization is another form of natural degradation (Lal, Hall & Miller, 1989) that limits crop production. From an ecological perspective, naturally saline soils are a soil condition requiring specific plant materials. Secondary salinization and alkalization, caused by irrigation, contributed to the abandonment of about 10 million ha of irrigated land each year (Szabolcs, 1987). In arid environments, dryland salinity occurs when water evaporates from the soil, leaving salts behind. Replacing perennial vegetation with shallow-rooted annual crops can cause dryland salinity. The lower water-use of these crops increases groundwater recharge and raises water tables (Ruprecht & Schofield, 1991; Schofield, 1992) (Figure 2.7). This

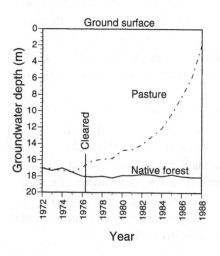

Figure 2.7. Changes in groundwater levels following clearing and pasture development after conversion of watershed from native forest to pasture in Lemon catchment, Western Australia. From Ruprecht & Schofield (1991). Used with the permission of Elsevier Science (*Journal of Hydrology*).

additional groundwater may then pick up additional salts and seep out at lower elevations in the landscape (Berg, Naney & Smith, 1991; Hobbs, Saunders & Arnold, 1993).

Secondary salinization has other causes. Salt-water intrusions in the Middle East caused salinization where excessive groundwater pumping allowed seawater to penetrate inland. Excessive irrigation from the Senegal River allowed seawater to move up-river and caused salinization (Thomas & Middleton, 1993). Reductions in the Nile River flow following construction of the Aswan High Dam created similar problems in the Nile delta (Kishk, 1986). Oil field activities such as the onsite disposal of drilling fluids and leakage from water lines containing salt water have degraded large areas. Oil and gas production activities caused soil salinity problems on 174 000 ha in Texas alone (McFarland, Ueckert & Hartmann, 1987).

Erosion

Soil erosion is the most common and damaging form of degradation since it ultimately degrades the physical, chemical, and biological properties of soils, and is irreversible. Both abiotic and biotic processes affect erosion. Soil erosion rates in North American piñon–juniper ecosystems are very sensitive to changes in biotic controls (ground cover). Where the soil erosion potential is high, erosion rates can rapidly change from low to high (Davenport *et al.*, 1998). Since erosion reduces the water-holding capacity of a soil, it increases the frequency of drought. This reduces plant production and increases death losses of individual plants and entire species. These damaged biotic components capture and retain fewer resources. This positive feedback degradation loop ultimately degrades energy capture, nutrient cycling, and hydrologic processes.

Natural erosion occurs under natural conditions of climate, vegetation, and landform. It is the result of natural geomorphic processes operating on landforms undamaged by humans. Accelerated erosion is more rapid than natural erosion and is the direct or indirect result of human activities. We must recognize the relative magnitude of natural erosional processes, but should concentrate our activities on reducing accelerated erosion. Although fluvial and eolian processes may operate on the same site, they usually occur at different times.

Erosion prediction models estimate erosion losses on specific sites and can compare alternative management strategies (Wischmeier & Smith, 1978). The most widely used erosion prediction models (universal soil loss equation and the wind erosion equation) are best suited to cultivated lands. The universal soil loss equation (USLE) is most useful for relative comparisons of management practices, since the soil erodibility, cover management, and supporting practice factors for erosion prediction are suspect (± 100%) under wildland conditions. The wind erosion equation (WEQ) uses information on erodibility, soil surface roughness, unsheltered length of eroding field and the vegetative cover to arrive at an estimate of wind erosion (Woodruff & Siddoway, 1965). Despite difficulties quantifying wildland erosion rates, these erosion prediction models are valuable tools that suggest useful repair strategies. The practical implications of the USLE and WEQ for repairing damaged primary processes suggest the value of strategies that (1) increase ground cover of vegetation and litter; (2) increase soil surface roughness; and (3) reduce the unobstructed length (fetch) of exposed soil.

WATER (FLUVIAL) EROSION

Water erosion has four distinct types: (1) interrill; (2) rill; (3) gully; and (4) streambank (Lal, 1990; Hudson, 1995). Interrill erosion is soil movement by rain splash and the transport of raindrop-detached soil by thin surface flow. Turbulence caused by raindrop impact also increases erosive capacity. There is no precise dividing line between rills and gullies. The most common description is that rill erosion creates small washes that are eliminated with normal cultivation. Gullies are so large and well established that farm implements cannot cross them. Streambank erosion is caused by the force of running water that undercuts stream and river banks (Lal, 1990).

On slopes, erosion begins with raindrop impacts on unconsolidated surface materials that loosen the soil, making it susceptible to detachment and erosion. Raindrop impact is the primary energy source for detaching unprotected soil. Rainsplash seals the soil surface and reduces the infiltration rate. When rainfall intensity exceeds the infiltration capacity of the soil, water collects on the surface and fills microtopographic depressions. When these depressions are filled, the excess water flows down the slope. Overland flow (runoff, surface flow, and sheet flow) and interrill erosion (work performed by rainsplash and overland flow) operate in a sheetlike fashion. Thin layers of water flow downslope and remove uniform layers of soil from slopes. These laminar water movements occur at moderate velocities, exert little force, and produce little erosion (Emmett, 1978). Rather than a uniform sheet movement, water develops flow concentrations over surface depressions as small as 1 mm (Bryan, 1979). These shallow streams and threads of water shift their position back and forth across the hillslope and over a number of storms remove relatively uniform layers of soil.

Rill processes are substantially more erosive than overland flow and are a function of hillslope length, depth of flow, shear stress, and critical discharge (Toy & Hadley, 1987). Rill erosion starts when the eroding force of the flow exceeds the ability of the soil particles to resist detachment. Flow depth and velocity are substantially greater where surface irregularities concentrate overland flow into rills. Once rills are established, the concentrated flow develops more detachment force and the rill formation process is enhanced. Rill development moves upslope as headcuts. Some rills develop rapidly and become more deeply incised than others. These 'master rills' become longer and deeper than their neighbors do. Occasionally, flows from adjacent rills break into master rills by eroding the boundary between them. As the rill flow becomes concentrated toward master rills, previously parallel rills develop a recognizable dendritic drainage pattern that accelerates rill development.

As rills coalesce, flow concentration and velocity increases until the more deeply incised rills become gullies. A gully is 'a channel or miniature valley cut by concentrated runoff but through which water commonly flows only during and immediately after heavy rains or during the melting of snow; it may be dendritic or branching or it may be linear, rather long, narrow, and of uniform width' (SCSA, 1982). Soil, hydrologic, landform and management factors cause gullies. They are most common in arid and semiarid regions with denuded landscapes and flash floods. Although the amount of sediment transported by interrill erosion probably exceeds erosion from within the gully by several orders of magnitude (Lal, 1992), gully erosion is more dramatic and far more difficult to control.

The likelihood of gully formation is increased by abrupt textural breaks in surface and subsurface horizons, subsurface flows (especially piping), high dispersible clay content and poor soil structure (Lal, 1992). Human-caused factors such as cultivation, deforestation, abusive grazing, road construction, footpaths, and engineered structures that concentrate runoff tend to hasten gully formation and accelerate its growth. This concentrated water moves soil, nutrients, and organic materials from the site. The deepening of rills and slumping of side slopes creates gullies. Slumping is caused by the sheering effect of the concentrated overland flow, increased water pressure within pores, and additional slumping caused by tunneling and pipeflow. Animal burrows near the gully often accelerate lateral gully extension (Lal, 1992).

In addition to interrill, rill, and gully erosion, subsurface waterflows cause erosion. Some of the precipitation that infiltrates into the soil moves downslope as 'interflow.' Overland flow erosional processes are more important in arid and semiarid regions. Interflow is of greater importance in humid regions. Two types of subsurface flows are important (Toy & Hadley, 1987). The first is matrix flow where water moves through granular or small structural pores. The second is pipe flow where water moves through larger subterranean channels or voids. Though not well understood, matrix flow is believed to account for a very small percentage of hillslope erosion. In contrast, pipeflow has great erosive force but is seldom apparent. These erosive forces continue to enlarge the pipe until the roof collapses and forms a gully.

WIND (EOLIAN) EROSION

Approximately 80% of the 3700 million ha of rangeland around the world are affected by wind erosion (UNEP, 1977). Wind erosion is greatest on fine soil components such as silt, clay, and organic matter. This causes some wind-eroded sites to have elevated levels of sand, gravel, and other coarse materials. Wind-blown particles are moved in three ways: (1) saltation, the bouncing of particles across the surface; (2) suspension; and (3) surface creep, the movement of larger particles caused by the pushing action of saltating particles striking the larger particles from behind. The rate of wind erosion depends on soil erodibility, surface roughness, climate, the unsheltered travel distance of wind across a field, and vegetative cover. Increasing soil roughness, reducing wind speed, and increasing the percentage of the soil covered by living vegetation or litter reduces wind erosion. Rough soil surfaces reduce wind erosion by disrupting airflows across its surface.

Assessing wildland processes

Ecosystems have unique combinations of processes that contribute to proper functioning. Intact ecosystems have many functions. However, we must begin by understanding the processes that control the flow of water, soil, nutrients, and organic materials within the landscape. The choice of processes to assess is determined on a site- and situation-specific basis (Meyer, 1997). What processes are most damaged? What additional understanding is most important to the design of effective repair strategies? What are the landscape management objectives? What caused the damage? What are our repair goals and against what standard should they be compared? How should we measure changes in selected criteria? What rate of repair should we expect? These questions seldom have clear-cut answers (Hobbs & Norton, 1996). Although these questions are often discussed (Cairns, 1989; Berger, 1991; Westman, 1991; Kondolf & Micheli, 1995; Hobbs & Norton, 1996), few generalizations have emerged.

Although numerous suggestions for assessing specific parameters are available (Table 2.1) (BLM, 1973; Le Houérou, 1984; Schaeffer, Herricks & Kerster, 1988; Costanza, 1992; Costanza, Norton & Haskell, 1992; Aronson *et al.*, 1993b; Aronson *et al.*, 1993a; BLM, 1993; NRC, 1994; Tongway, 1994; Tongway, 1995; Whisenant *et al.*, 1995; Tongway & Ludwig, 1997b), no single assessment parameter is appropriate for all situations. While some have argued that reference ecosystems are necessary (Aronson, Dhillon & Floc'h, 1995), others contend they lead to unattainable goals (Pickett & Parker, 1994; Hobbs & Norton, 1996). Although setting reference conditions as narrowly defined goals is unrealistic, using reference ecosystems of similar landform/soil/biota/climate is useful and practical (Hobbs & Norton, 1996). For each of the identified attributes, the repaired ecosystem could be compared with its reference ecosystem by calculating similarity indices (Berger, 1991; Westman, 1991; Kondolf, 1995; Kondolf & Micheli, 1995). Comparisons with equivalent, less damaged areas make it easier to identify processes important to the long-term persistence of natural ecosystems (Yates, Hobbs & Bell, 1994).

An alternative approach (Hobbs & Norton, 1996) is based on the structural, compositional, and functional measurements originally proposed for 'ecosystem health' assessments (Costanza *et al.*, 1992). This approach compares current conditions against the estimated range of natural variability of any relevant parameter (Hobbs & Norton, 1996). This assessment strategy could be expanded by developing ecological reference templates that define a limited range of functional and structural states (Allen, 1994). Then the current state of essential parameters for a site is compared with the estimated range of natural variability for the same parameters (Caraher & Knapp, 1995).

Each process is evaluated at multiple spatial scales, from microtopography to landform and watershed features, with clear goals and time scales for each parameter. Since ecosystem processes and structures show properties at multiple scales, the scale at which we make our observations is critical (Lewis *et al.*, 1996). Regardless of what scale we focus on, it is essential to examine larger scales for an understanding of context and smaller scales for insight into underlying mechanisms (Lewis *et al.*, 1996). Since processes damaged at one scale influence all smaller scales, large scale problems are not adequately assessed or repaired with data or actions limited to much smaller scales.

Unfortunately, wildland repair funding is seldom sufficient to intensively assess all possible parameters. Many of these assessments require relatively intensive, site-specific sampling, and laboratory analyses. Since indiscriminate analyses easily overwhelm wildland repair budgets, we must identify priority processes for additional assessment. The degree, scale, and economics of wildland repair demand pragmatic approaches. Fortunately, some visual assessments of primary processes are rapid, inexpensive, and readily apparent.

A pragmatic approach assigns priority to assessment parameters that address the conservation of limiting resources and the functioning of essential primary processes. Since the condition of the soil surface conveys useful information on essential hydrologic and nutrient cycling processes, damage assessments should begin with soil surface features and associated hydrologic processes (Tongway & Ludwig, 1997a; Tongway & Ludwig, 1997b). Hydrologic assessments include infiltration, runoff, hydraulic conductivity, deep drainage, water-table depth, transpiration, vegetation type, and plant cover. Several visual assessments of hydrology and nutrient cycling provide additional focal points for developing practical and effective repair strategies.

Table 2.1. *Process-related assessment parameters that have been suggested for wildland repair activities*
These process-related parameters should be compared with the same parameter in similar, undisturbed sites.

Assessment parameter	Importance	Reference
Soil organic matter	Influences soil structure, soil stability, hydrologic and nutrient cycling processes	Aronson et al. 1993a
Soil surface conditions (crusting, erosion features, litter, vegetation and microbiotic cover, microtopography)	Provides information on erosion (past and future susceptibility) in addition to the state of hydrologic and nutrient cycling processes. See Tables 2.2 and 2.3	BLM (1973); Aronson et al. (1993a); Aronson et al. (1993b); NRC (1994); Tongway (1994); Tongway (1995)
Rain use efficiency (annual, aboveground primary production / annual rainfall)	Assessing the health and productivity of arid ecosystems	Le Houérou (1984)
Coefficient of rainfall infiltration (amount of water moving into the soil)	Indicator of surface-soil conditions	Aronson et al. (1993a)
Maximum available water reserves	Of particular importance where precipitation is irregular or where soil erosion losses have occurred. Integrates depth and water retention capacity of soil	Aronson et al. (1993a)
Length of water availability	Provides information useful in determining suitable species and degree of degradation	Aronson et al. (1993a)
Depth and quality of water table	Indicator of seriously altered hydrologic processes operating at large scales	Whisenant (1993); Hobbs & Norton (1996)
Cation exchange capacity	Indicator of capacity of soil to retain cations	Aronson et al. (1993a)

Nutrient pools	Indicator of degree of nutrient depletion and provides information on selecting suitable species	Bradshaw (1983); Schaeffer et al. (1988)
Cycling indices (ratio of the amount of energy or element recycled to the amount of energy moving straight through the system)	Integrates numerous soil, vegetative, hydrologic, and climatic variables	Aronson et al. (1993a)
Microbial biomass, microbial respiration, net mineralization	Assessment of microbial activity	Hart, August & West (1989); Santruckova (1992)
Total carbon and microbial carbon	Assessment of microbial recuperation of soil as an indicator of ecosystem recovery	Insam & Haselwandter (1989); Ruzek (1994)
Metabolic quotient (ratio of microbial respiration to microbial biomass)	Used to assess the state of microbial recovery of soil as an indicator of ecosystem recovery	Insam & Haselwandter (1989)
Soil dehydrogenase activity, ATP and ergosterol	Three-dimensional ordination to assess the restoration potential and progress of reclaimed soils	Bentham et al. (1992)

Soil stability and hydrologic functioning

Soil surface conditions and surface roughness are useful indicators of hydrologic processes, nutrient retention, nutrient cycling, and biological processes.

SOIL SURFACE CONDITIONS

Soil condition assessments provide insight into stability (ability to withstand erosive forces), hydrologic processes (infiltration and runoff), and nutrient cycling (BLM, 1973; Aronson *et al.*, 1993a; NRC, 1994). As an example, interrill erosion on Texas rangelands was inversely related to grass and litter abundance (Thurow, Blackburn & Taylor, 1988). Not only does vegetation affect hydrologic processes, bare soil surfaces have very different properties and induce different types of hydrologic behavior. Soil stability and hydrologic functioning are assessed with information on texture, structure, soil erosion (fluvial and eolian processes), soil deposition, and the infiltration or capture of precipitation.

Soil texture provides a preliminary estimate of soil permeability and water retention ability of the entire soil profile. The hydraulic conductivity of soils increases one order of magnitude for each step in the textural series, clay to loam to sand (Loomis & Connor, 1992). The texture of subsurface soil layers is important, since clay pans or fragipans reduce potential productivity. Clay pans or fragipans are impervious to water and severely restrict the soil's storage capacity and rooting depth. Very coarse textured soils do not need much vegetation or litter to have relatively high infiltration rates. Thus, coarse sands with damaged vegetation may still have functional hydrologic processes.

Soil structure conveys much about soil hydrologic function and ability to resist erosion. Soil structure is used to assess the structural quality of the topsoil based on the air capacity and available water (Hall *et al.*, 1979). An ideal soil structure has abundant pores and fissures over 0.1 mm diameter that permit free root growth, oxygen diffusion, and water movement. It will also have pores smaller than 0.05 mm that hold water against the force of gravity. Rapid infiltration prevents water accumulation on the surface. Subsoil compaction should be absent, and erosion should not reduce potential rooting depth (Tivy, 1990). This structural quality rating system is useful, but still does not

46

consider the importance of a range of pore sizes to allow for ease of root penetration, free drainage, and adequate water storage.

Clay soils in the tropics have different physical and chemical properties than clay soils in temperate soils. Despite clay contents as high as 85%, tropical soils are often quite friable, because they are aggregated into fine structural units (Young, 1974). This microaggregation is caused by the cementing properties of free iron oxides, which are common in tropical soils. The physical effect is more like sandy soils than clay soils, since they are freely permeable and easily penetrated by crop roots. However, they have a low nutrient-holding capacity and are very susceptible to nutrient losses by leaching. In tropical soils, nutrient-holding properties of organic matter are proportionately more important than in temperate soils. Because of rapid decomposition rates, organic matter contents are lower in tropical soils and concentrated more near the surface than in temperate soils (Tivy, 1990).

Heavy equipment or even livestock compact soils, particularly when wet. Restricting equipment access on wet soils reduces damage and subsequent repair requirements (Davies, Younger & Chapman, 1992). Compacted soils are major problems of old fields, abandoned roads, and mined soils (Sopper, 1992). Although bulk density is the most commonly used measure of compaction, packing density is more reliable because it adjusts for the influence of clay content (Coppin & Stiles, 1995). Packing density (D_p) is calculated as:

$$D_P = D_{DB} + (0.009 \cdot C_\%) \tag{2.1}$$

where D_{DB} is the dry bulk density (mg m^{-3}) and $C_\%$ is the percentage of clay content.

Soil surface stability is assessed as the degree and nature of: (1) soil movement; (2) surface rock and/or litter; (3) pedestaled plants (Figure 2.8) or rocks; (4) flow patterns; and (5) rills and gullies (Table 2.2). A more comprehensive approach uses eleven soil surface attributes (Table 2.3) to assess damage to ecological processes (Tongway, 1994). The objective of this assessment procedure is to estimate resistance to erosion (stability), hydrologic responses (infiltration versus runoff), and organic matter cycling efficiency (nutrient cycling). Although developed for Western Australia, this process-oriented approach is applicable, with minor changes, to most wildlands.

Live vegetation or litter, an intact mineral soil crust, surface gravel, or the presence of a vigorous microbiotic crust increase the stability of

Table 2.2. *Erosion-related surface soil characteristics and assessment classes*

Characteristic	Class 1	Class 2	Class 3	Class 4	Class 5
Soil movement	Subsoil exposed on much of the area; may have embryonic dunes and/or wind scoured depressions	Soil and debris deposited against minor obstructions	Moderate movement of soil particles has occurred	Some movement of soil particles has occurred	No visual evidence of soil movement
Surface rock and/or litter	Very little remaining; if present, surface rock or fragments exhibit some movement and accumulation of smaller fragments behind obstacles	Extreme movement; many large deposits against obstacles; surface rocks exhibit movement; smaller fragments accumulate behind obstacles	Moderate movement; fragments deposited against obstacles, fragments have a poorly developed distribution pattern	May show slight movement; if present, coarse fragments have a truncated appearance or spotty distribution caused by wind or water	Accumulation in place; if present, the distribution of fragments shows no movement caused by wind or water
Pedestaling	Most rocks and plants pedestaled and roots are exposed	Many rocks and plants pedestaled and roots are exposed	Rocks and plants pedestaled in flow patterns	Slight pedestaling in flow patterns	No visual evidence of pedestaling
Flow patterns	Flow patterns numerous, readily noticeable; may have large barren fan deposits	Flow patterns contain silt, sand deposits and alluvial fans	Well defined, small and few with intermittent deposits	Deposition of particles may be in evidence	No visual evidence of flow patterns

Rills and gullies	May be present at depths of 8–15 cm and at intervals of less than 13 cm; sharply incised gullies cover most of the area, with 50% actively eroding	Rills 1–15 cm deep at 150 cm intervals; gullies numerous and well developed; active erosion on 10–50% of their lengths or a few well-developed gullies with active erosion along more than 50% of their length	Rills 1–15 cm deep in exposed places at about 300-cm intervals; gullies well developed, with active erosion along less than 10% of their length with vegetation present	Few infrequent rills in evidence at distances of over 300 cm; evidence of gullies with little bed or slope erosion; some vegetation is present on slopes	No visual evidence of rills; may be present in stable condition, but with vegetation on channel bed and side slopes

Source: BLM (1973).

Table 2.3. *Soil surface assessment features for determining soil condition*

Assessment characteristic	Class					
	1	2	3	4	5	6
Soil cover – raindrop interception	<2%	1–2%	2–5%	5–15%	15–50%	>50%
Soil cover – overland flow obstruction	0%	<2%	2–5%	5–15%	15–50%	>50%
Crust brokenness	extensively broken	moderately broken	slightly broken	intact		
Microbiotic crust cover	<1%	1–10%	10–50%	>50%		
Erosion features	extensive	moderate	slight	nil		
Eroded materials	extensive	moderate	slight	nil		
Litter cover	<1%	1–10%	10–25%	25–50%	50–100%	100% and several cm thick
Soil microtopography	smooth <3 mm deep	few shallow depressions 3–8 mm	deeper depressions 8–15 mm	deep, extensive 15–25 mm	sink holes >25 mm	
Surface nature	loose-sandy, over noncoherent sand	Crust is easily broken with finger pressure, and is brittle. Subcrust is noncoherent	Crust is moderately hard (needs plastic or metal tool to break), but	Crust is very hard (needs metal tool to break surface), but is brittle,	Crust shows some flexibility when pressed with pen or finger pressure,	

	Very unstable	Unstable	Moderately stable			Stable	Very stable
Slake test	Very unstable. Fragment collapses completely in < 2 seconds with a myriad air bubbles into shapeless mass	Unstable. Fragment substantially collapses over about 5 seconds, a thin surface crust remains, but > 50% of the subcrust material slumps to an amorphous mass	Moderately stable. Surface crust remains intact, some slumping of subcrust material, but < 50%	brittle, breaking into amorphous fragments or powder. Subcrust is coherent	breaking into amorphous fragments or powder. Subcrust is hard and coherent	Stable. Whole fragment remains intact over periods of 1 hr or more	or surface is self-mulching clay. Subcrust is coherent or strong crumb structure
Soil texture	Silty clay to heavy clay	Sandy clay loam to sandy clay	Sandy clay loam to silt loam	Sandy loam to silt loam	Sandy to clayey sand		

Notes:

Soil cover (raindrop interception) – is the projected cover of perennial grasses and perennial shrubs to a height of 0.5 m. It also includes rocks, sticks, and other relatively immovable and long-lived objects that protect the soil from raindrop impact. Soft annual herbage and woody vegetation (> 0.5 m) are not included in this assessment.

Soil cover (overland flow obstruction) – is the projected cover of long-lived objects that obstruct overland flows. This slows flow rate, reduces losses and captures resources from other parts of the landscape.

Table 2.3. (cont.)

Crust brokenness – concerns the thin layers of fine-textured soil particles over the soil. This assessment is used to determine the degree that the surface crusts are broken, loosely attached and susceptible to erosion. More broken crusts are more susceptible to erosion. Soils without natural crusts (self-mulching surfaces and loose sands are excluded).

Microbiotic crust cover – percentage of the soil surface covered by microbiotic crusts. Healthy microbiotic crusts increase soil surface stability.

Erosion features – are visible signs of active soil loss. This includes incised channels, rills, sheet erosion and pedestaling caused by soil loss.

Eroded materials – is used to estimate the degree to which materials are being eroded from one place in the landscape to another. These features include unconsolidated depositions that are easily remobilized and lost again. These materials are lost from the source site, but may improve the productive potentials on the sink site. Excludes litter.

Litter cover – is used as an estimate of the availability of organic materials for decomposition and nutrient cycling. It is recommended that the source of the litter fraction be identified: locally produced and deposited or produced and transported from another source.

Soil microtopography – is used as an assessment of the ability of the site to detain water. Soil depressions reduce runoff, capture soil and nutrients while increasing infiltration into the soil.

Surface nature – is used to evaluate the robustness of the surface, the resistance to disturbance, and its resistance to raindrop erosion or other physical disturbances.

The slake test – is an evaluation of crust stability when immersed into rain water. Stable crusts maintain their cohesion when wet and are more resistant to erosion.

Soil texture – provides an estimate of soil permeability beneath the surface crust.

Source: Tongway (1994, 1995).

Figure 2.8. Seriously pedestaled switchgrass plant with numerous exposed roots indicates rapid soil loss and highly damaged hydrologic processes. It is important to differentiate between mounded plants that have trapped soil and organic materials and plants, like this one, that are pedestaled because the soil has been eroded from around them.

soil surfaces. In climates with the potential for complete coverage (high productivity and/or slow decomposition), soil cover is an effective assessment of soil condition. Although well-covered and aggregated soil surfaces are most desirable, arid and semiarid regions may not be able to completely cover the soil. Since bare soils in these environments cannot maintain the structure necessary to prevent surface crusting, it is useful to assess the relative stability of bare soil surfaces. The bare surface assessment techniques used in Table 2.3 are easy to use and interpret (Tongway, 1994). Another assessment methodology was developed that describes nine functionally distinct surface crusts of semiarid West Africa and relates them to unique management problems and potentials (Casenave & Valentin, 1992).

Intact surface crusts with a thin layer of fine-textured soil suggest less damage and greater resistance to additional erosion (Table 2.3). This attribute (crust brokenness) assesses the degree to which the surface crust is broken, loose, or loosely attached and susceptible to erosion (Tongway, 1994). Bare soils in good condition have a smooth crust that conforms to gentle undulations in the soil surface. Smooth unbroken soil crusts erode less than crusts with partially broken fragments.

The nature of the soil surface (surface nature in Table 2.3) indicates its resistance to trampling or raindrop impact. Soils with fragile crusts are easily damaged and eroded. Flexible crusts have fine roots and/or fungal hyphae holding the soil particles together and greater biological activity. Very hard crusts resist detachment, but also have lower infiltration rates and diminished organic matter levels. Surface nature is determined on dry soils by examining crust flexibility, brittleness, and the coherence of the subcrust soil (Table 2.3) (Tongway, 1994, 1995).

Another soil surface stability assessment is determined by immersing soil crust fragments in a beaker of rainwater and observing the response of fragments over time (Table 2.3). Stable crusts maintain their cohesion better when wet and are more able to resist the erosive effects of flowing water (Tongway, 1994). Fragments that collapse with a myriad air bubbles into a shapeless mass in less than 2 seconds are unstable and less beneficial. Unstable fragments substantially collapse over 5 seconds, retaining a thin surface crust, but greater than 50% of the subcrust material slumps into an amorphous mass. Stable fragments remain intact for at least 1 hour.

Interrill erosion, rills, channels, gullies, and pedestaled plants indicate soil loss (Tables 2.2 and 2.3). Interrill erosion and scour erosion by wind can remove or reduce the depth of the A-horizon. The loss of the A-horizon, and the presence of rills, gullies, scoured soils, and pedestals are indicators of soil loss (Table 2.2). Rill and gully developments are practical indicators of soil erosion, reduced infiltration, and nutrient loss. The depth, distribution, and development of a branching pattern of rills and gullies suggest the degree of erosion and runoff (Table 2.2). Pedestaled plant height and the depth of root exposure provide estimates of recent soil loss.

The accumulation of eroded materials (1) around plants or in small basins, (2) as sediment in alluvial fans, gullies, streams, or lakes, or (3) as dunes, indicates erosion occurred elsewhere (Tables 2.2, 2.3). Deposits range from small accumulations around plants or other obstructions to large fan-shaped deposits. These soil deposits, in and around plant bases, should not be confused with pedestals. Soil accumulations around plants were captured from erosion occurring elsewhere in the landscape. Pedestaled plants indicate onsite erosion.

Animals have important influences on soil aeration, drainage, and friability. Earthworms, termites, ants, and beetles improve aeration and friability through their burrowing activities. Earthworms, termites, and certain vertebrates bring soil from lower strata to the surface, which increases nutrient availability to plants. An Australian study found that ants can bring $841 \text{ g m}^{-2} \text{ yr}^{-1}$ of soil to the surface while earthworms turn up only $133 \text{ g m}^{-2} \text{ yr}^{-1}$ (Humphreys, 1981). In Australia, ants are bioindicators of ecosystem repair after mining (Andersen & Sparling, 1997). This is supported by the correlation of aboveground ant activity with belowground decomposition processes. Cultivation and livestock grazing also affect soil structure. Seven years after stopping cultivation and removing sheep, densities of larger soil organisms were dramatically higher (Abbott, Parker & Sills, 1979). This is of special significance because larger soil organisms increase water permeability and soil porosity.

SURFACE ROUGHNESS

Surface roughness influences wind and water flow rates, which affects erosion and other hydrologic processes. It operates at several spatial scales and contributes to both abiotic and biotic resource retention mechanisms. Abiotic surface roughness is assessed at scales that consider depressions in the soil (microcatchments, furrows, plowing, hoof prints, etc.) or aboveground obstructions like rocks and gravel. It is viewed at larger scales by assessing the influence of relative landform position. The biotic influences on surface roughness include vegetation (density, cover, height, stiffness) and nonliving organic obstructions such as herbaceous litter and woody debris on the soil surface.

Flow velocities are reduced in dense, uniform vegetation. Open, clustered vegetation is less effective and allows localized erosion where flow velocities increase between vegetation clumps (Styczen & Morgan, 1995). When flow depths are shallow relative to the vegetation, plants maintain their height and a high degree of roughness. With increased flow rates, plant stems begin to oscillate, causing more disturbance to the flow and increasing resistance to flow (Morgan & Rickson, 1995c). When plants are submerged by deep flows or flattened by rapid flows, resistance decreases by as much as an order of magnitude. Thus, the relative stiffness (resistance to horizontal flows) of plants is an important feature. Living and dead organic materials protect the surface from raindrop impact and reduce flow velocity. Organic materials work indirectly by providing cementing substances that bind soil aggregates and maintain porosity and infiltration (Satterlund & Adams, 1992). Inorganic objects, like gravel and rocks, use their large mass to resist erosion.

The coefficient of friction is an effective measure of surface roughness since it includes the effects of raindrop impact, flow concentrations, tillage effects, litter, rocks, and the erosion and transport of sediment (Engman, 1986). The coefficient of friction is Manning's coefficient 'n', which was developed to assess channel stability under various flow rates. Manning's coefficient is not directly measured, but is described by its effect on the flow rate of water (Morgan & Rickson, 1995c).

In channels, resistance to erosion is expressed relative to its maximum permissible velocity (Satterlund & Adams, 1992), which is the greatest velocity a particular channel can withstand without erosion. It is calculated with the Manning equation:

$$V = \frac{R^{\frac{2}{3}} S^{\frac{1}{2}}}{n} \qquad (2.2)$$

where V is the mean velocity (m s^{-1}), n is the Manning coefficient of roughness (dimensionless), R is the hydraulic radius (m), and S is the slope (m m^{-1}). The hydraulic radius, R, is a channel shape factor that depends on channel cross-section dimensions and depth of flow. The channel shape factor is calculated as:

$$R = \frac{A}{WP} \qquad (2.3)$$

where A is the cross-sectional area of flowing water (m^{-2}) and WP is the length (m) of the wetted perimeter and its containing channel (measured at right angles to the direction of flow) (Satterlund & Adams, 1992). In shallow flows, as in surface runoff, R is equal to the flow depth (m) (Styczen & Morgan, 1995), so its value for surface runoff is very small, usually 0.001 or less (Satterlund & Adams, 1992). The effects of raindrop impact, which is much of the detaching force, complicate the determination of safe velocities (Satterlund & Adams, 1992). However, it still provides a relative comparison among treatment alternatives.

THRESHOLD WIND EROSION VELOCITY

Soil particles move when wind exceeds the friction threshold velocity (FTV) for a particular soil surface. Since the length of exposed soil needed to reach load-capacity depends on the erodibility of each soil, maximum transport loads occur at lower wind speeds and in shorter distances on more erodible soils. Winds of 18 m s^{-1} (at 10 m above the surface) acquire maximum transport loads within 55 m on a structureless fine sand, but need more than 1500 m on a cloddy medium-textured soil (Chepil & Woodruff, 1963).

The size distribution of saltating soil grains is a function of the size distributions of loose aggregates of the eroding surface (Gillette *et al.*, 1980). FTV increases at about the half-power of the size of the mode of the aggregate size distribution. Thus, soil surfaces with more large aggregates have higher FTVs, meaning they require more wind to move soil particles (Gillette *et al.*, 1980). Nonerodible elements, such as gravel, stones, and vegetation increase threshold velocities for a particular soil, making it more difficult to erode. Thus, in many wildlands, wind erosion continues across the soil surface until enough noneroding elements are uncovered to protect remaining erodible grains (Middleton, 1990). The protective effect of these noneroding elements is most apparent on wind stable surfaces such as 'stone pavements' or 'desert pavements', which occur in sparsely vegetated areas like hot deserts. Soil surfaces with microbiotic crusts tolerate higher wind speeds compared with bare surfaces. Physical damage to microbiotic crusts significantly reduced FTVs by 73–90%. Unfortunately, microbiotic crusts damage easily and may require 20 years to recover (Belnap & Gillette, 1997).

RAIN USE EFFICIENCY

Rain use efficiency (RUE) is a simple, indirect estimate of some aspects of hydrologic function. RUE is the quotient of annual, aboveground primary production divided by the annual rainfall and it has been recommended as a useful tool for assessing the health and productivity of arid ecosystems (Le Houérou, 1984). RUE generally decreases with increasing aridity and potential evapotransporation but is surprisingly consistent, in similar management situations, throughout the world. Soil condition features such as permeability, texture, depth, water storage capacity, and fertility status all affect RUE. In arid regions, the highest RUEs occur on soils that store most of the water from scarce rains. In natural vegetation of arid and semiarid ecosystems, RUE is usually between 1.0 and 6.0 (100 to 2000 kg ha^{-1} yr^{-1}), but may be much lower on degraded sites and much higher in pristine ecosystems. Since RUE integrates climate, vegetation, and soil condition, it approximates both site condition and potential (Le Houérou, 1984).

PROPER FUNCTIONING CONDITION OF RIPARIAN WETLANDS

The functioning of riparian wetland systems should be assessed with a process-oriented approach (Table 2.4) that emphasizes the site's capacity to conserve its resources (BLM, 1993). This assessment refers to the status of vegetation, geomorphic, and hydrologic development, along with the degree of structural integrity exhibited by the riparian-wetland area. Healthy riparian wetland areas are in dynamic equilibrium with streamflow forces that shape the stream channel. Healthy riparian wetlands adjust, with limited channel and vegetation perturbation, by altering their form and slope to handle increased runoff. Complete functional assessments require consideration of conditions within the entire watershed. The watershed affects the quality, amount, and stability of downstream resources by controlling the production of sediments and nutrients, influencing streamflow, and by modifying the

Table 2.4. *Elements of proper functioning condition for use in assessing riparian-wetlands*

Riparian-wetland areas are functioning properly when adequate vegetation, landform, or large woody debris is available to:

(1) dissipate stream energy associated with high water flows, thereby reducing erosion and improving water quality;

(2) filter sediment, capture bedload, and aid floodplain development;

(3) improve floodwater retention and groundwater recharge;

(4) develop root masses that stabilize streambanks against cutting action;

(5) develop diverse ponding and channel characteristics to provide the habitat and the water depth, duration, and temperature necessary for fish production, waterfowl breeding, and other uses;

(6) support greater biodiversity.

Source: BLM (1993).

distribution of chemicals throughout the area. Proper function does not require the presence of any particular species since the degree of structural integrity is the primary objective of this assessment (BLM, 1993). However, this might be used in conjunction with other objectives that require the presence of certain species or species assemblages.

Nutrient cycling

While the amount of nutrients in wildlands is important, it is less important than nutrient-related processes. Degraded wildlands are often deficient in the major nutrients. We can increase nutrient levels with fertilizer applications, but the benefits of this strategy are often short-lived. Ultimately, nutrient limitations must be addressed through the restoration of organisms that recycle nutrients from litter to the soil. Rather than focus on the amount of nutrients (structure), we must repair damaged nutrient cycling processes (function). There are practical and ecological reasons for this functional emphasis. From a practical perspective, it is very difficult to determine what nutrient levels are best for natural communities. Wildland economies seldom support continuing fertilization. From an ecological perspective, agronomic assessments of nutrient requirements are inappropriate for wildlands because (1) nutrient availability varies among ecosystems;

(2) nutrient-rich sites are inherently more responsive to nutrient additions than nutrient poor sites; and (3) nutrient demand and plant response change with age and successional status (Chapin, Vitousek & Cleve, 1986).

NUTRIENT AVAILABILITY IN ECOSYSTEMS

We must determine whether the site is degraded or inherently infertile. Assessing nutrient limitations is not simple. Low nutrient levels do not always indicate problems and high nutrient levels are not always desirable. Many plant communities evolved under low fertility and are poorly equipped to compete under enriched conditions. Low nutrient availability favors native prairie species over most weeds and non-natives (Biondini & Redente, 1986; Hobbs & Atkins, 1988; Huenneke *et al.*, 1990; Wilson & Tilman, 1991; Wilson & Gerry, 1995). Many species adapted to infertile sites are unable to take advantage of elevated fertility. The 'paradox of enrichment' states that high amounts of nutrients, such as nitrogen, favor some species to the detriment of most other species (Rosenweig, 1987). Thus, elevated fertility inhibits species richness (Marrs, 1993). Elevated N availability inhibits succession in both mesic (Tilman, 1984; Tilman, 1987; Aerts & Berendse, 1988; Carson & Barrett, 1988) and semiarid ecosystems (McLendon & Redente, 1991; Pashke *et al.*, 1996). Restoring native vegetation to these sites requires less fertility (see Chapter 3 for more discussion).

COMMUNITY RESPONSE TO FERTILIZATION

Greater productivity after fertilization does not indicate nutrients are limiting community development. We fertilize crops to increase productivity, but we control competing plants to prevent natural changes from occurring. Plants from infertile areas have mechanisms for conserving nutrients (see Chapter 5). Under nutrient-limiting conditions, many plant species have low tissue nutrient concentrations. They also translocate a high percentage of their nutrients from senescing leaves (Starchurski & Zimka, 1975; Shaver & Melillo, 1984), thus producing litter with low nutrient concentrations. In addition, leaves from plants in nutrient-poor sites decompose and release nutrients slowly compared with plant materials from nutrient-rich sites. These traits conserve nutrients by reducing the rate of nutrient cycling. However, the same traits make plants from infertile ecosystems less able to take advantage of added nutrients.

Nutrient subsidies are less available on nutrient-poor sites compared with the same nutrient addition on nutrient-rich sites. Moderate nutrient additions (i.e., a 50% increase in annual nutrient flux) are less available on infertile sites than on fertile sites because of immobilization processes (Chapin *et al.*, 1986). Decomposers in nutrient-poor sites are energy limited due to the low decomposability of the litter (Flanagan & Cleve, 1983). Phenolic compounds and lignin, which are abundant in the litter from nutrient-poor sites, reduce decomposition rates and nutrient release by inhibiting the activities of microorganisms (Aber & Melillo, 1982). Nutrient-poor sites also produce litter with high carbon to nutrient ratios. This increases nutrient immobilization in the microbial biomass and slows nutrient release (Bosatta & Staaf, 1982).

Not only is nutrient availability important, the form of each mineral resource is critical. For example, the form of available nitrogen changes during prairie development, with nitrate-nitrogen (NO_3-N) dominating early in prairie development. Nitrates are easily leached from soils, and plants expend more energy to absorb nitrate since it must be reduced to NH_3. Mature prairies produce more biomass per unit N with NH_4-N than with NO_3-N (Pickett, Collins & Armesto, 1987b). Although mature prairie species (i.e., *Schizachyrium scoparium* and *Andropogon gerardii*) can establish soon after disturbance, they often perform poorly during early successional development. This slow initial growth is probably due to low NH_4-N availability in disturbed areas (Pickett *et al.*, 1987b).

NUTRIENTS AND SUCCESSION

Nutrient demand and plant response changes with age and successional status. Mid- to late-seral species are adapted to low N availability, whereas early seral species are better adapted to high N availability. Thus, decreased N availability during secondary succession shifts the competitive advantage to mid- and late-seral species. Conversely, increased N availability increases the competitive advantage of early seral species and inhibits succession. Ecosystems probably conserve more nutrients as they mature (Odum, 1969), but this relationship is more complicated. Intermediate aged ecosystems are better at retaining nutrients than very young or very old ecosystems (Vitousek &

Reiners, 1975). It seems likely that plant production is generally more limited by N on young soils and by P on old soils (Walker & Syers, 1976; Vitousek & Farrington, 1997). Ecosystems receive nutrients from many sources; and they lose nutrients to a variety of sinks. Ecosystems capture and retain more nutrients as they develop. Biomass accumulation provides for additional nutrient capture and storage. Thus, nutrient gains will begin to exceed nutrient losses as indicated by the accumulation of additional biomass. Thus, the ecosystems' ability to retain nutrients is greater than in the earlier developmental sequences. However, ecosystems cannot continue to capture more nutrients than they lose. At dynamic equilibrium, ecosystem production is believed to equal zero and net production will be zero (Odum, 1969). When this occurs, net nutrient output is roughly equal to net nutrient input.

Visual assessments of nutrient cycling

Solar energy drives the development of fully functional wildlands. The organic materials produced by plants are essential to the development and maintenance of damaged wildlands. Although little definitive information is available, indicators such as the degree of fragmentation in the distribution of plants, litter, roots, and photosynthetic period are useful starting points for understanding nutrient cycles and energy flow (NRC, 1994). Nutrient capture is most efficient when plants growing at different seasons share the rooting zone. Rooting systems that exploit more soil volume have greater access to water and nutrients. The presence of actively growing plants during the entire growing season suggest more effective use of available nutrients (NRC, 1994).

Simple, visual assessments of the soil surface provide additional useful information about nutrient cycling (Tables 2.2 and 2.3). Litter abundance suggests the availability of organic materials for decomposition and nutrient cycling (Tongway, 1994). The amount and distribution of organic materials indicates which areas are losing organic materials and which areas are accumulating them. Litter abundance and incorporation into the soil provide rough approximations of nutrient cycling processes. The relative contribution of litter to nutrient cycling on a particular site can be divided into three categories: (1) litter that is loosely strewn across the landscape has no value; (2) litter in intimate contact with the soil surface has a slight value; and (3) litter that partially or wholly covers the soil surface has extensive value (Tongway, 1994). Is the litter of local origin or has it been transported from another site?

Perennial, nitrogen-fixing legumes are essential components of many ecosystems (Jenkins, Virginia & Jarrell, 1987; Jarrell & Virginia, 1990). The presence and abundance of nitrogen-fixing species indicate nitrogen is being added to the system. Several studies documented greater nitrogen availability in soils with nitrogen-fixing pioneers (Lawrence *et al.*, 1967; Vasek & Lund, 1980; Hirose & Tateno, 1984; Vitousek *et al.*, 1987). A study of natural vegetation dynamics following the retreat of glaciers at Glacier Bay, Alaska, suggested the importance of a nitrogen-fixing component in successional development (Crocker & Major, 1955). However, nitrogen-fixing plants do not always strictly facilitate succession. At least two studies found nitrogen fixers may also inhibit successional development (Walker & Chapin, 1986; Morris & Wood, 1989).

3

Repairing damaged primary processes

Introduction

Traditionally, wildland repair programs emphasized soil modifications to suit the desired species, rather than using adapted species that improve soil conditions. The objective was to rapidly alter soil conditions to meet the requirements of desired plant species by changing the soil's physical, chemical and biological attributes (Table 3.1). Fertilization, liming, and other subsidy-based approaches are effective and have well-developed methodologies (Schaller & Sutton, 1978; Bradshaw & Chadwick, 1980; Bradshaw, 1983; Lal & Stewart, 1992; Munshower, 1994). Few wildlands have the productive potential to finance those subsidies, particularly when they must continue indefinitely. Thus, contemporary approaches are often so expensive that severely degraded wildlands are more likely to be abandoned than repaired (Harrison, 1992). This is particularly true in arid and semiarid environments, where the risks are greater and potential returns are lower.

Soil textbooks often include descriptions of the 'ideal' soil. This ideal soil has a loamy texture, granular structure, good fertility, and organic matter content and contains approximately half solid matter and half pore space (e.g., Brady, 1990). The solid mineral component comprises about 45% of the soil volume with the remaining 5% being organic matter. At optimum soil-water levels this 'ideal' soil contains equal proportions of air and water in the pore spaces and has physical, chemical, and biological features contributing to plant growth. Most soils are less than optimum, due to inherent properties (parent material, climate, or age) or accelerated (anthropogenic) degradation. While desirable and occasionally achievable as an agricultural objective, this ideal soil is usually neither possible nor realistic as a wildland repair objective. This 'ideal soil' may not even be desirable, since natural communities are adapted to the entire range of soil conditions.

Rather than attempting to create an ideal soil, a more pragmatic goal is to develop healthy soils. Soils are healthy by comparison with their potential, rather than by comparison with some unattainable standard. Healthy soils maintain the integrity of essential processes. We must design management strategies that repair those processes. Addressing surface soil problems repairs many dysfunctional hydrologic or nutrient-cycling processes. Where soil degradation is minor, the obvious standard of comparison is a similar undisturbed soil. However, where degradation has been severe (e.g., significant soil erosion), the standard of comparison is less apparent. Rather than using direct comparisons among soil characteristics to determine repair strategies, I suggest an approach that seeks to repair damaged processes.

Repaired ecosystems have more biotic control over resource flows. Although a variety of initial treatments will initiate these repair processes, sustained improvement requires adapted plants. Damaged wildlands require species that not only grow under existing conditions, they must initiate autogenic repair processes that continue to improve ecosystem functioning. Plants increase resource retention, which produces more plants. Those plants capture still more resources, initiating a positive feedback loop of autogenic repair. Since we are seldom able to change landform, the most exciting wildland repair opportunities lie in small-scale physical modifications. Well-chosen surface soil modifications can initiate and direct autogenic processes that continue to improve an expanding portion of the landscape.

It is neither possible, nor is it my objective, to cover all soil problems or repair strategies. The primary emphasis is on soil surface conditions and associated processes (crusting, erosion, runoff, and infiltration). A secondary emphasis is on problems associated with nutrient cycling, salinization, gully erosion, and compacted soils. While it is necessary to assess and repair site-specific damage, we cannot restrict our repair efforts to site-specific problems. Wildland repair programs that design solutions at spatial scales that are too small may experience unanticipated problems. Small-scale actions seldom repair basin-scale problems. In this situation, small-scale solutions address the symptoms of

Table 3.1. *Soil problems and potential strategies for treating wildland soils*

Nature of the problem		Initial treatment	Long-term treatment
Physical			
Structure	too compact	rip. scarify, cultivate, incorporate organic matter or apply soil conditioner	increase vegetation, root biomass and litter
	too open	compact or cover with fine material	increase vegetation, root biomass and litter
Stability	unstable	soil stabilizer or mulch	reshape slope, establish slope stabilizing vegetation
Moisture	too wet	drain	use vegetation adapted to flooded conditions; use vegetation with high transpiration rate to lower water table
	too dry	organic mulch, pitting, microcatchments, contour furrowing, obstructions (e.g., logs or rocks) on soil surface to retain water and organic materials	establish vegetation with structural and functional diversity
Chemical			
Macronutrients	insufficient quantity or availability	fertilizer, use species adapted to infertile environments	establish species or symbiotic associations capable of fixing atmospheric nitrogen; establish vegetation with mycorrhizal infections; continue fertilizer and lime subsidies; use woody vegetation to retain nutrients on-site

Micronutrients	insufficient quantity or availability	fertilizer, use species adapted to infertile environments	increase vegetative and microbial diversity; increase woody vegetation to retain nutrients on-site
pH	too basic	pyritic waste or organic matter	weathering, acidifying vegetation
	too acidic	lime, adapted vegetation	cation pumping vegetation
Toxicity	heavy metals	organic mulch or tolerant vegetation	inert covering or tolerant vegetation
	sodium	weathering or irrigate with gypsum	tolerant vegetation
Biological			
Soil organic matter	low percentage of soil	organic matter amendments	vegetation
Soil organisms	low diversity and activity	reduce physical and chemical limitations of the soil, organic matter amendments, reintroduce missing organisms if their requirements are met	reduce physical and chemical limitations of the soil; vegetation

Source: Adapted with modifications from Bradshaw (1983).

larger problems, without affecting the cause. Thus, wildland repair programs should attempt to repair damaged processes at the largest scale that is damaged (Rabeni & Sowa, 1996).

Improving soil surface conditions

Increasing the stability and infiltration rate of the soil surface initiates the repair of damaged processes through a positive feedback system that improves both soil and vegetation. This positive feedback loop operates in reverse to the degradation processes (see Figure 2.6). In the long-term, processes controlled by soil surface conditions are only repaired and maintained by increasing plant production and protecting the soil surface with plant litter or living vegetation. However, short-term treatments are used to 'jump-start' soil-repairing processes (Whisenant, 1995).

Initial soil surface treatments involve at least one of four general approaches. One, increase the roughness of the soil surface with pits, contour furrows, basins, ripping or chiseling. Two, add aboveground obstructions such as logs, rocks, woody debris, herbaceous litter, or man-made erosion control products. Three, use soil conditioners to rapidly improve surface structure. Four, encourage the development of microbiotic crusts on degraded soil surfaces. The first two techniques do not attempt rapid soil structural improvement and are less frequently used on sandy soils. Ponding water on the surface increases infiltration and captures more soil, nutrients, seed, and other organic materials from fluvial and eolian flows. The third is a short-term strategy that uses soil conditioners to improve the surface soil structure. The fourth strategy encourages the development of microbiotic crusts. This is an intriguing possibility, with few specific recommendations.

Increase surface soil roughness

Because many degraded wildlands occur in arid or semiarid climates, limited water availability often restricts their recovery. Therefore, surface soil treatments often focus on reducing runoff and erosion by retaining precipitation near where it falls. The wind erosion equation and universal soil loss equation confirm the value

Figure 3.1. Recently created microcatchments used to reestablish woody plants on severely crusted soils in Niger. Where previously little water moved into these soils, these microcatchments held enough water to allow the establishment of large shrubs that began autogenic development. Photograph courtesy of Thomas L. Thurow.

of soil surface roughness, which suggests the importance of (1) increasing vegetative cover; (2) increasing surface roughness; and (3) reducing the distance of unobstructed soil. The possibility of achieving each of these objectives is increased, directly or indirectly, with microcatchments (Figures 3.1), pits (Figure 3.2), contour furrows (Figure 3.3), basins, ripping, or chiseling. These strategies are effective because microroughness influences infiltration rates of dry, fine-textured soils (Dixon & Peterson, 1971). Natural depressions in the soil accumulate water, nutrients and organic matter that aid establishment of woody plants in arid and semiarid regions (Ahmed, 1986; Kennenni & Maarel, 1990). Man-made depressions in the soil concentrate scarce resources and initiate soil repairing processes (Whisenant *et al.*, 1995). Surface modifications on slowly permeable Montana soils increased Precipitation Use Efficiency (PUE) more than 100% (Wight & Siddoway, 1972). The same treatments increased PUE about 20% on sites with high infiltration capacities. These processes facilitated vegetative development that

Figure 3.2. Pits created in crusted soils of old field near Pecos, Texas, with farm equipment (furrow-diker) to hold water long enough to facilitate plant establishment that will continue site development.

Figure 3.3. Contour furrows in Shaanxi Province, People's Republic of China, created to establish woody plants on very steep hillsides.

continues to capture an increasing percentage of the organic matter and nutrients flowing across the landscape in wind and water. These improvements contribute to vegetation change and influence the trajectory of vegetation development.

CONTROLLING WIND EROSION WITH ROUGHER SOIL SURFACES

Controlling wind erosion involves the application of two basic principles: (1) reduce the wind velocity near the soil surface; and (2) increase the resistance of the soil surface to wind drag (Lal, 1990). Wind velocities near the soil surface can be reduced with (1) afforestation; (2) temporary cover crops or preparatory crops; (3) plant residue, rocks, or logs on the surface; or (4) shelterbelts. Management practices that improve soil structure and conserve soil moisture increase the soil's resistance to wind drag. Tillage and soil management practices, prior to planting, can reduce erosion and improve success. Surface ridges produced by tillage influence erosion rates. Their impact is determined by the height and lateral frequency of ridges, furrow shape, orientation relative to wind direction, and the proportion of erodible to nonerodible grains (Middleton, 1990). Tillage ridges are more effective when oriented at right angles to erosive winds. Furrows that are parallel to the wind may increase soil loss by increasing the scouring influence through the furrows. Tillage practices reduce wind erosion by slowing saltation and surface creep if they use crop residue (preparatory crops) or produce a very rough cloddy seedbed with furrows perpendicular to the prevailing wind direction (Lal, 1990; Potter, Zobeck & Hagan, 1990). Practices that improve soil aggregation by adding soil organic matter, mulch, or soil conditioners reduce wind erosion. Since moist soil is less susceptible to wind erosion, cultural practices that conserve water are particularly helpful. These practices include mulches, preparatory crops, cover crops, or even irrigation where possible.

Severe wind erosion in the Mohave Desert north of Los Angeles created serious respiratory problems, poor visibility, impassable roads, and excessive sand deposition on nearby homes (Spitzer, 1993). The Los Angeles County Fire Department reduced these problems with emergency revegetation of over 1000 ha of abandoned farmland. Ripping created 20-cm deep furrows perpendicular to the prevailing winds. Grass seed were drilled into the furrow bottoms and the entire

71

area was overseeded with adapted shrubs. After two growing seasons, woody plants were abundant on 95% of the area and no severe wind erosion events occurred in the planting area (Spitzer, 1993).

Add aboveground obstructions

Many wildland repair programs emphasize engineered structures to concentrate and accelerate runoff. This emphasis on diversion works, graded banks, waterways, and similar projects ignores the fundamental problem, the reduced infiltration and water-holding capacity that produces excessive runoff from damaged wildlands. The most effective wildland strategies promote the retention and use of water where it falls and do not allow the water to develop enough velocity to cause erosion. Surface covers that reduce the amount and velocity of surface or channel flows are most effective. However, on the most barren sites, aboveground obstructions are necessary before plants can establish.

TYPES OF ABOVEGROUND OBSTRUCTIONS

A variety of aboveground obstructions provide enough initial stability to establish vegetation that provides lasting benefits. They also capture and concentrate resources that increase plant establishment and accelerate vegetative development. Aboveground obstructions reduce the flow rate of wind and water across the soil surface. Obstructions may capture water, nutrients, and organic materials and increase the infiltration rate. In Niger, mulching barren crusted soils increased soil moisture, seed capture, development of ground cover, and the germination of woody species after a single rainy season (Chase & Boudouresque, 1987). Contour rows or rocks on gentle slopes reduced water movement, increased infiltration, and reduce erosion (Figure 3.4). Logs (Figure 3.5), felled trees, and brush piles provide similar benefits.

PLANTS AS OBSTACLES TO FLUVIAL FLOWS

Plants are self-sustaining obstructions that increase surface roughness. They reduce flow velocities, increase soil stability, and increase the amount of water infiltrating into the soil. Compare alternative treatments for increasing surface roughness with Manning's roughness

Figure 3.4. Rows of rocks placed along slope contour in Niger as aboveground obstacles to trap water, soil, nutrients and organic materials. Although the area was not seeded, the water and soil trapped above the rocks led to the natural establishment of herbaceous vegetation that continues to improve the site.

Figure 3.5. This log was placed on bare, recently seeded ground near College Station, Texas, to capture soil, water, nutrients, and organic matter flows. Two months after seeding the effect of this aboveground obstacle on plant establishment is apparent.

coefficient (Equation 2.2). The potential range of Manning's roughness coefficient is large, and the differences are important. For example, doubling the roughness coefficient decreases flow velocity 34% and increases water depth 50% (Styczen & Morgan, 1995). Suggested values for Manning's roughness coefficient (Table 3.2) provide a starting point for evaluating alternative repair strategies.

Manning's roughness coefficient is a highly dynamic term that changes through the growing season. Although vegetation height is correlated with reduced flow velocities, that relationship is weak, because some plants bend over during strong water flows (Watts & Watts, 1990). These differences lead to the notion that the rigidity of individual species and flow depths relative to plant height should be considered by estimating Manning's n as a function of the deflected roughness height k (m) (Kouwen & Li, 1980) in the following way:

$$n = \frac{y^{\frac{1}{6}}}{(8g)^{0.5}(a + b \log(y/k))} \tag{3.1}$$

where y is flow depth (m), g is a gravity term (m s^{-2}), and the values of a and b depend on the ratio of shear velocity to critical shear velocity. The value of k is a function of the stiffness index (MEI):

$$k = 0.14h \left[\frac{\left(\frac{MEI}{\lambda y S}\right)^{0.25}}{h} \right]^{1.59} \tag{3.2}$$

where h is the deflected roughness height of the vegetation (m) and λ is the unit weight of water. The MEI is the flexual rigidity of vegetation elements per unit area. M is the number of elements per square meter, E is the modulus of elasticity of the vegetative material (measured in newtons (N) m), and I is the second moment of the cross-sectional area of the stems (m^4). Multiplying these components yields a value of MEI in N m^2. The vegetation begins to bend and flatten at the critical shear velocity (m s^{-1}), which is defined by (u^*_{crit}) = 0.028 + 6.33 (MEI)^2 (Kouwen & Li, 1980). Stiffness index (MEI) values, for several species (Kouwen & Li, 1980; Morgan & Rickson, 1995c) (Table 3.3), provide relative comparisons for estimating stiffness values for other species.

We can compare the importance of flexual rigidity to roughness assessments by contrasting a turf-forming grass with an erect, stiff bunchgrass. In channels, Bermuda grass (*Cynodon dactylon*) protects

Table 3.2. *Suggested values for Manning's coefficient (n), which is a measure of surface roughness*

Manning coefficient indicates the relative protection from flowing water with larger numbers providing greater protection from flowing water. Although usually used for channels, it may be used for surface flows. Manning's n values should be considered dynamic since they vary seasonally and may be reduced by an order of magnitude after they are flattened by strong, deep flows.

Ground cover or treatment	Residue (g m²)	Suggested value	Range
Concrete or asphalt		0.011	0.010–0.013
Bare sand		0.010	0.010–0.016
Graveled surface		0.020	0.012–0.030
Bare clay loam (eroded)		0.020	0.012–0.033
Packed clay		0.030	–
Fallow, no residue		0.050	0.006–0.160
Shortgrass prairie		0.150	0.100–0.200
Dense shrubs and forest litter		0.400	0.330–0.475
Dense grass		0.240	0.170–0.300
Bermuda grass		0.410	0.300–0.480
Light turf		0.200	0.165–0.225
Dense sod		0.350	0.325–0.400
Bluegrass (*Poa pratensis*) sod		0.450	0.390–0.630
Chisel plow	< 60	0.070	0.006–0.170
	60–250	0.180	0.070–0.340
	250–750	0.300	0.190–0.470
	> 750	0.400	0.340–0.460
Disk / harrow	< 60	0.080	0.008–0.410
	60–250	0.160	0.100–0.250
	250–750	0.250	0.140–0.530
	> 750	0.300	–
No tillage	< 60	0.040	0.030–0.070
	60–250	0.070	0.010–0.130
	250–750	0.300	0.160–0.470
Moldboard plow		0.060	0.020–0.100
Coulter		0.100	0.050–0.130

Source: Modified from Engman (1986), Satterlund & Adams (1992), Styczen & Morgan (1995).

Table 3.3. *Suggested values for the stiffness index (MEI; N m²), for selected vegetation*

Vegetation type	*MEI* value
Medicago sativa, green and uncut	2.9–6.2
Cynodon dactylon, green and long	1.5–47.4
Cynodon dactylon, green and short	0.03–0.6
Buchloë dactyloides, green and uncut	0.03–0.7
Bouteloua gracilis, green and uncut	4.2–6.0
Eragrostis curvula, green and long	3.1–15.4
Poa pratensis, green and short	0.01–0.2
Lespedeza striata, green and short	0.005
Lespedeza striata, green and long	0.02–3.0
Lespedeza cuneata, green and short	0.015
Lespedeza cuneata, green and long	6.3–15.9
Pennisetum clandestinum, green and long	35.0–57.0
Pennisetum clandestinum, green and short	0.14–0.21
Chloris gayana, green and long	96.0–212.0

Source: (Adapted from Morgan & Rickson, 1995c).

the soil surface from raindrop impact and erosive water flows. However, it has little influence on water velocity. Nor does it capture much of the soil, nutrients, or organic materials in the water, because the flexible stems lay over under relatively low water flows. In contrast, the stiff stems of switchgrass (*Panicum virgatum*) remain erect under strong water flows, reducing water velocity (Kemper *et al.*, 1992), and trapping scarce resources (Figure 3.6) Chapter 5 discusses other plant attributes that stabilize sites and repair damaged processes.

PLANTS AS OBSTACLES TO EOLIAN FLOWS

Plants affect wind erosion in at least five ways (Morgan, 1995). First, plant foliage reduces wind speed by exerting a drag on airflows. Plant biomass, projected foliage area facing the wind, leaf area density, leaf orientation, and leaf shape all influence a plant's ability to reduce wind erosion. Vegetation influences on wind speed vary seasonally, particularly with deciduous species. Second, the foliage traps moving sediments. This not only reduces erosion, it has important implications for nutrient dynamics on the site since eolian dust has substantially more

Figure 3.6. Single switchgrass (*Panicum virgatum*) plant transplanted into actively eroding channel near College Station, Texas, trapped organic materials (upslope) and about 13 cm of soil (downslope) within 60 days.

nutrients than degraded wildland soils (Drees, Manu & Wilding, 1993). Third, vegetative cover protects the soil surface. Fourth, plant root systems increase the resistance of the soil to erosional processes. Fifth, vegetation influences soil moisture through uptake, transpiration, and microenvironmental modifications.

Our understanding of how plants alter wind erosion is not sufficient to reliably simulate their effects on wildlands. However, we can effectively design wildland repair programs to reduce wind erosion. We do that by increasing vegetative cover, plant height, and soil surface roughness while reducing the length of unprotected soil. Recent studies indicated that plant area index and canopy cover are highly correlated with the transport capacity of wind and provide safe indicators of soil protection (Armbrust & Bilbro, 1997). In general, plants that are taller, finer-leafed, and have large surface areas will be most effective in reducing wind erosion (Middleton, 1990).

PREPARATORY CROPS

Obstacles that slow the wind and capture wind-blown soil particles reduce wind erosion (Floret, Floc'h & Pontanier, 1990). Mechanical and chemical stabilization techniques are expensive and have a limited effectiveness period. Lasting stabilization requires vegetative cover. However, the conversion from bare ground to perennial is difficult because moving sand particles kill emerging seedlings. Preparatory crops provide a practical approach to the transition from cropping to perennial vegetation. Preparatory crops are annual plants that stabilize sites and improve seedbed conditions before planting the permanent vegetation.

Preparatory crops (usually an annual species) provide a relatively stable and safe environment for planting the perennial vegetation. Grain sorghum (*Sorghum* spp.) is the most common preparatory crop in sandy, semiarid west Texas. It is recommended to (1) grow the grain sorghum as a preparatory crop, and harvest the grain with normal farming practices; (2) sell the grain to partially offset repair costs; (3) leave the sorghum stalks standing to reduce wind erosion; (4) next spring, drill perennial grasses into the standing sorghum stalks. The stalks reduce erosion before planting, and after planting, and increase perennial plant establishment. Erect residues of the preparatory crop are more effective than horizontal residues because vertical residues absorb more wind energy (Siddoway, Chepil & Armbrust, 1965). The height, diameter, and number of stalks determine the effectiveness of standing residue, because they determine the silhouette area through which winds pass (Bilbro & Fryear, 1994).

SHELTERBELTS

Afforestation, with adapted species, reduces erosion by decreasing wind speed, protecting the soil surface, and increasing litter. Shelterbelts are most effective and most practical when the trees provide additional benefits to the ecosystem and/or local economy (e.g., amelioration of microenvironment, nitrogen fixation, fuel-wood, fodder, or wildlife habitat). Shelterbelts are highly effective and widely used to reduce windspeed, erosion, and evapotranspiration in areas susceptible to wind erosion. They should be planted perpendicular to prevailing winds. Taller trees provide protection for greater distances than shorter trees, but may also require rows of shorter trees or shrubs to fill in the lower level gaps. The shelter effect is determined with the following equation:

$$S = 1 - \frac{V}{V_f} = \exp \frac{a-3}{a} \qquad (3.3)$$

where S is the shelter effect, V is the wind velocity at distance a from the belt, V_f is the wind speed in the open, and exp is the logarithm (Lal, 1990).

In semiarid parts of Australia, planting 5% of the land area to shelterbelts, timberbelts, or tree blocks reduced windspeed by 30–50% and reduced soil loss by up to 80% (Bird *et al.*, 1992). In Sudan, shelterbelts that reduced sand encroachment had common attributes (Mohammed, Stigter & Adam, 1996) that led to several recommendations. These include planting shelterbelts with several wide rows placed perpendicular to prevailing winds and placing tall trees in the center, with many rows of dense shrubs on the outside of the shelterbelts. Shelterbelts with trees and dense shrubs are relatively impermeable to wind and have few gaps (Mohammed *et al.*, 1996). The most effective shelterbelt species grew rapidly, had long life spans, were tolerant of existing stresses, and provided valuable products for local inhabitants.

Use soil conditioners

Soil crusting results from the physical disintegration of soil aggregates (Coughlin, Fox & Hughes, 1973) and the chemical dispersion and movement of clay particles (Agassi, Shainberg & Morin, 1981). Soil particles are washed into the voids where alternating wetting and drying cycles create the crust (Chen *et al.*, 1980). Soil crusts reduce germination, decrease infiltration rate (Herbel *et al.*, 1973), and are a major cause of grass seeding mortality (Rubio *et al.*, 1989). Soil conditioners are economically feasible on some cultivated crops, but their expense restricts their wildland applications to critical sites. Polyacrylamides (PAM) are synthetic polymers that bind soil particles and reduce crusting, thus increasing pore space and infiltration. Three grass species emerged sooner from a hard crusted fine-textured soil, following PAM soil treatments that reduced crust formation (Rubio *et al.*, 1989). In Kenya, field applications of polyacrylamides to untilled crusted soils slightly increased infiltration rates and reduced soil loss rates over a period of six weeks (Fox & Bryan, 1992). Applying polyacrylamides to tilled soils produced dramatic, but short-lived effects. Incorporating polyacrylamides at 0.01% (of the soil weight) reduced runoff, decreased soil loss, and inhibited crust development. However, the effects only lasted several weeks and the cost was estimated at approximately US $190 ha^{-1} (1985 figures) (Fox & Bryan, 1992).

Initiate microbiotic crust development

Returning microbiotic crusts to degraded soils is an attractive strategy without an established methodology. The primary barriers are a lack of suitable inocula and high water requirements (Knutsen & Meeting, 1991). Water limitations are serious in arid and semiarid regions, but less of a problem in humid regions. Algal culture techniques are available, but economic restrictions prevent their use on wildlands. When added to sandy soils through center pivot sprinklers, mass-cultured *Clamydomonas* and *Asterococcus* species (Chlorophyceae) significantly improved soil aggregation (Meeting, 1990). Experience with algal inoculation of rice fields and microalgal soil conditioners suggests the potential for accelerated development of algal populations under certain conditions. In Utah, slurries of mature microbiotic crusts were an effective inocula for dry soils (St. Clair, Johansen & Webb, 1986). Although existing microbiotic crusts were destroyed to mix these slurries, the study supported the idea that semiarid soils can be inoculated with microbiotic species (Belnap, 1993). It may be possible to inoculate soil binding and N$_2$ fixing microalgae onto semiarid and desert soils (Meeting, 1990). However, practical strategies for establishing microbiotic crusts on degraded soils, using cultured organisms, are not available.

Increasing resource retention

The extent of nutrient loss from ecosystems is one assessment of ecosystem stability (Jackson, Selvidge & Ausmus, 1978). Sustainable ecosystems balance nutrient inputs with nutrient losses, but degraded wildlands must increase nutrient pools by capturing more and/or losing fewer nutrients. Increasing nutrient pools in degraded ecosystems is a serious economic challenge. After implementing soil surface treatments that contribute to greater resource capture, our next objective is to close the nutrient cycle by increasing nutrient retention.

Repair strategies that capture limiting resources increase primary production. This additional vegetation improves resource retention by

(1) increasing soil organic matter; (2) increasing water and nutrient holding capacities; and (3) improving soil structure. These changes initiate a positive feedback system of autogenic repair that continues to increase resource retention. Resource retention in damaged ecosystems is increased with repair strategies that (1) use vegetation compatible with the nutrient cycling regime; (2) repair or replace damaged soil biotic processes; and (3) add organic materials to the soil.

Match vegetation with the nutrient regime

Low fertility and acidic soils create serious obstacles to prevailing wildland repair approaches, particularly where management objectives require rapid recovery. We are concerned with long-term sustainability and are not attempting to recreate the 'ideal soil.' Thus, our perspective on fertility and nutrient cycling differs from the prevailing agronomic or mineland reclamation approaches. Neither the scales of operation nor the economic realities of damaged wildlands permit the widespread use of fertilizer and organic amendments. Sustaining high nutrient-demanding species with fertilizer inputs is a matter of economics. Trying to maintain successional or high nutrient-requiring species in low nutrient soils results in disappointment (Burrows, 1991). Where we cannot afford fertilizer, we match plant materials to the fertility regime. Rather than subsidizing species with high nutrient requirements, we should repair processes that capture and retain nutrients, and establish vegetation that is more compatible with the fertility regime.

NITROGEN

Since the amount of nitrogen (N) usually changes more than other nutrients during ecosystem development (Marrs *et al.*, 1983), repairing nitrogen cycling on severely degraded sites is important (Leopold & Wali, 1992). Repairing wildland N cycling, after soil removal or serious alteration (e.g., mineland or severely eroded wildlands), requires more efficient nutrient retention mechanisms or artificial nutrient subsidies (Bradshaw, 1983). Natural approaches require more time, since they depend on vegetative development and organic matter accumulation. Frequently, meeting land-use objectives within a reasonable time frame requires nutrient subsidies. Unfortunately, nitrogen subsidies are counterproductive in some situations, regardless of cost.

The development of nitrogen capital is one of the most important factors in ecosystem development after mining and soil removal (Bradshaw & Chadwick, 1980; Roberts *et al.*, 1981; Bradshaw, 1983; Palmer, 1992). Kaolin mine spoils near Cornwall, England, develop serious N deficiencies that prevent ecosystem development (Bradshaw *et al.*, 1975). Although a single nitrogen application can provide enough N for one year's growth, it does not provide for subsequent years. Long-term N requires the mineralization of organic N. Reclamation studies suggest that a nitrogen capital of $750–1000$ kg ha^{-1} is necessary to reestablish a functional nitrogen cycle (Roberts *et al.*, 1981; Bloomfield, Handley & Bradshaw, 1982; Bradshaw, 1983). Severely depleted sites might require annual nitrogen applications for five to ten years, if no other major sources of nitrogen are available, and the goal is a soil N capital of 750 kg ha^{-1} (Bloomfield *et al.*, 1982). In England, the total N pool necessary to maintain an annual release of 100 kg N ha^{-1} and an organic decomposition rate of $1/16$ per year was estimated to be 1600 kg N ha^{-1} (Bradshaw, 1983). Although many developed ecosystems have more N than this, it is a reasonable approximation for new ecosystems (Bradshaw, 1983).

A Colorado study that examined abandoned fields after 53 years of natural recovery found that C and N pools only recovered if perennial bunchgrass, such as *Bouteloua gracilis*, were present (Burke, Laurenroth & Coffin, 1995). Abandoned fields dominated by annuals did not accumulate significant levels of soil nutrients (Vinton & Burke, 1995). Thus, recovery of these abandoned, shortgrass–steppe fields depended on the establishment of perennial grasses, which are necessary for the accumulation and persistence of soil organic matter.

The amount of available N, relative to total N, is more important to ecosystem development than the size of the N pool (Skiffington & Bradshaw, 1981). High levels of available N inhibit development in many wildland ecosystems. In California, undisturbed sites with native vegetation had less available N, but more total N (Zink *et al.*, 1995). Exotic annuals dominated the adjacent, disturbed sites with more available N (compared with total N). The disturbed site produced litter that was easily degraded. Litter from native vegetation degraded at a slower rate, so it accumulated more litter, and created conditions that favored native species (Zink *et al.*, 1995).

Elevated N availability also inhibits succession and the intended development of seeded wildlands. Although this relationship is well documented in both mesic (Tilman, 1984; Tilman, 1987; Aerts & Berendse, 1988; Carson & Barrett, 1988; Marrs & Gough, 1989; Marrs, 1993; Clarke, 1997; Snow & Marrs, 1997; Stevenson, Ward & Pywell, 1997) and semiarid ecosystems (Hobbs & Atkins, 1988; Huenneke *et al.*, 1990; McLendon & Redente, 1991; Pashke *et al.*, 1996), the implications of elevated N levels for wildland repair efforts are often ignored. Mid- and late-seral species are adapted to conditions of low N availability, whereas early seral species are better adapted to conditions of high N availability. Thus, a decrease in N availability during secondary succession provides the competitive advantage to mid- and late-seral species. Conversely, increased N availability increases the competitive advantage of early seral species and inhibits successional development. Another study found antelope bitterbrush (*Purshia tridentata*) seedling growth was significantly increased following carbon (sucrose) applications that reduced the growth of herbaceous annuals (Young, Clements & Blank, 1997). The sucrose applications inhibited weedy annuals by reducing available N levels, which allowed shrub seedlings to develop without competition.

Single applications of large amounts of inorganic N are wasted in low organic matter soils with a low N retention capacity (Sopper, 1992). Minelands use annual fertilizer applications, but this is unrealistic in most wildland situations. One-time applications of 40 000–50 000 kg sewage sludge ha^{-1} are effective, since it provides about 1500 kg organic N ha^{-1} and 625 kg P ha^{-1} (Bradshaw, 1983). The reclamation success achieved with sludge is probably due to three factors related to its organic content: (1) the N content is in a slowly available organic form; (2) the high organic C content provides an immediate energy source for soil microbes; and (3) sludge organic matter improves soil physical conditions resulting from soil removal and compaction (Sopper, 1992).

Ecosystem response to fertilization is variable (Berg, 1980), creating controversy over its value in wildlands. Fertilization increases the growth of aggressive species and can contribute to stand deterioration. Overfertilization produces plant biomass amounts that easily exceed the decomposition capacity of immature soils (e.g., minespoils or severely eroded wildland soils). This additional litter immobilizes nutrients and disrupts nutrient cycling. Nitrogen fertilizers reduce the competitive advantage of legumes while benefiting the more nitrogen responsive species. Fertilization reduces species diversity by favoring ruderal or weedy species at the expense of species that dominate more highly developed ecosystems.

It is occasionally desirable to reduce the amount of available N in the soil (Morgan, 1994; Zink *et al.*, 1995; Clarke, 1997). In England, nutrient stripping and acidification are often necessary for the successful restoration of heathland vegetation (Smith, Webb & Clarke, 1991; Aerts *et al.*, 1995; Clarke, 1997; Snow & Marrs, 1997). Nutrient reductions are accomplished by encouraging leaching, removing vegetation (hay removal), by incorporating organic materials that tie up nutrients (Smith *et al.*, 1991; Clarke, 1997; Snow & Marrs, 1997), or even topsoil removal (Aerts *et al.*, 1995). Adding sawdust (or other material with a high C/N ratio) to soils in Manitoba stimulated microbial activity that tied up large amounts of N; reduced N availability and reduced *Poa pratensis* establishment (Morgan, 1994). However, adding sawdust in Saskatchewan to favor *Andropogon gerardii* over *Agropyron cristatum* and *Bromus inermus* only created more bare ground (Wilson & Gerry, 1995).

An alternative to large and/or frequent fertilizer applications is the use of nitrogen-fixing symbiotic relationships that provide a continuing, low-level N source. Nitrogen fixation benefits the plant and bacterial partners (Heichel, 1985). The plant host provides a hospitable environment, and nutrition, for the bacterial partner. The plant provides sugars, produced by photosynthesis, and other nutrients to the bacteria. The bacteria develop inside specialized root structures called nodules. These bacteria use sugars from host plants as an energy source to convert atmospheric N_2 gas to ammonium ions that plants use for amino acid production and protein synthesis. This symbiotic relationship is important because it supplies nitrogen to host plants and additional nitrogen to other plants. Studies on the reclamation of China clay wastes in England suggested legumes contributed to the development of a self-sustaining ecosystem (Dancer, Handley & Bradshaw, 1977). Legumes are important, not only because of the amount of N they contribute, but because they supply N in a relatively steady way through the growing season (Palmer & Chadwick, 1985; Palmer *et al.*, 1986). However, the N-contributions of legumes may be quite low without sufficient phosphorus in the soil (Palmer & Iverson, 1983).

Woody legumes and other woody species with nitrogen-fixing symbiotic relationships can contribute significant amounts of nitrogen to disturbed landscapes (Jeffries, Bradshaw & Putwain, 1981; Bethlenfalvay & Dakessian, 1984; Dawson, 1986; Reddell, Diem & Dommergues, 1991; Prat, 1992; Zitzer, Archer & Boutton, 1996). Well-adapted woody legumes added 190 kg N ha^{-1} yr^{-1} to degraded soils in Brazil (Franco & Defaria, 1997). Black locust (*Robinia pseudoacacia*) dominated clear-cut sites in the southern Appalachians of the eastern US, and was partly responsible for conserving nutrient pools after disturbance. Black locust is grown on minelands in the eastern US for its N-fixing ability. Total stand N increased 30–75 kg ha^{-1} yr^{-1} in aggrading, mixed hardwood stands in the southern Appalachians (Leopold & Wali, 1992).

The percentage of N and organic carbon in Sonoran desert soils increased under shrubs (Barth & Klemmedson, 1978; Cox, Parker & Stroelein, 1984). Total N, NO_3-N, organic carbon, $NaHCO_3$-extractable P, and saturation extract K were significantly higher beneath mesquite, while Na^+ and Cl^- ions were significantly higher away from the mesquite canopy (Virginia & Jarrell, 1983). Surface concentrations of nutrients were greater under big sagebrush (*Artemisia tridentata*) compared with soils between the shrubs and grass-influenced soils (Doescher, Miller & Winward, 1984).

SOIL PH

Natural processes, rather than human activities, create most acid soils. Humid environments create acid soils through erosion, base leaching, organic matter oxidation, pyrite oxidation, or acid deposition. Adding lime increases the availability of many nutrients, increases biological activity, and decreases heavy metal toxicity. Most soil organisms benefit from liming very acid soils, since bacteria and actinomycetes are more active near neutral pH. Liming is widely used in mineland reclamation, but long-term control of acidity is very difficult (Leopold & Wali, 1992) and lasting benefits are rare in wildlands. More soluble materials (Na_2CO_3, Na and Ca silicates, and gypsum ($CaSO_4·2H_2O$)) are more desirable for some situations. Rocky soils complicate the deep incorporation of lime.

Deep-rooted species obtain nutrients that shallow-rooted plants cannot get. This ability not only increases their survival and growth, it improves the habitat and facilitates the subsequent establishment of additional species. A good example is the use of 'cation pumpers' on acid soils to obtain Ca and Mg from deep in the soil to raise the surface soil pH (Zinke & Crocler, 1962; Alban, 1982; Kilsgaard, Greene & Stafford, 1987; Choi & Wali, 1995). Studies on acidic minespoils indicate that a pH of 3.7 to 4.0 is the lowest at which acid-tolerant species can be established without site amelioration (Vogel, 1984). Species that tolerate very acid soils are uncommon, but fortunately, these extreme pH levels are less common on wildland soils. Biological methods that rely on cation pumping species are able to raise the pH of acidified soils. The expansive root systems of some robust grasses (Choi & Wali, 1995) and woody plants (Zinke & Crocler, 1962; Alban, 1982; Kilsgaard *et al.*, 1987) are particularly useful, since they remove cations from lower soil horizons and concentrate them in the surface soil.

Black locust and aspen (*Populus tremuloides*) or hybrid poplar (*Populus* spp.), planted on acid mine spoils raised the pH of the surface soil. They also increased soil organic matter and soil nitrogen levels (Alban, 1982). Then, other species established through natural or artificial seeding methods. Cedars and related species of Cupressaceae and Taxodiaceae actively concentrate calcium in the soil under their canopies (Zinke & Crocler, 1962; Kilsgaard *et al.*, 1987). Their litter contains high concentrations of basic cations that reduce soil acidity. Aspen and white spruce (*Picea glauca*) with red pine (*Pinus resinosa*) and jack pine (*Pinus banksiana*) improved soil conditions on two sites in northern Minnesota. Forty years after planting, spruce and aspen stands had moved enough cations from lower soil horizons to the forest floor (Alban, 1982) to create a nearly neutral forest floor. Pine trees had the opposite effect; they acidified the forest floor.

NUTRIENT CYCLING

Although natural selection in low-fertility sites selects for attributes that conserve nutrients, we often do just the opposite. The prevailing paradigm for repairing degraded wildlands often involves the selection and breeding of plant materials for increased productivity and forage quality. Although this approach works well on fertile sites, it can create inherently unstable conditions on infertile sites. Plant materials reinforce patterns of nutrient availability because genotypes adapted to low-nutrient systems grow slowly, use nutrients conservatively, and produce herbage that is less attractive to herbivores and decomposer organisms (Grime & Hunt, 1975; Poorter & Remkes, 1990; Poorter, Remkes & Lambers, 1990; Aerts & Peijl, 1993).

Genotypes from nutrient-poor sites have several traits that slow nutrient cycling. Phenolics and lignin, common in litter from nutrient-poor sites, reduce decomposition and nutrient release rates by inhibiting microorganisms. They have low nutrient concentrations and translocate a high proportion of nutrients from senescing leaves. This low-nutrient litter decomposes and releases nutrients more slowly than litter from nutrient-rich sites. Since these plant traits increase litter accumulations, they are usually beneficial to degraded wildlands.

Plant materials selected for wildland repair grow rapidly and have high forage quality. These attributes accelerate nutrient cycling. Simulation modeling and empirical data indicate that productive, high-nutrient genotypes initially dominate nutrient-poor sites. However, this dominance is short-lived. The biomass of nutrient-conserving species often exceeds that of more productive species after a few years (Figure 3.7).

In an Australian study (Johnson & Tothill, 1985), nonnative species (*Cenchrus ciliaris, Panicum antidotale*, and *Chloris gayana*) performed very well following brush clearing, by exploiting nitrogen reserves released by the soil disturbance during brush removal. Over time, soil nitrogen levels declined and the pastures were less productive. This degeneration is normal and we should learn to anticipate its occurrence and manage accordingly (Myers & Robbins, 1991). That management requires that we use less-nitrogen demanding species, add nitrogen, or accept lower livestock numbers to delay the inevitable decline. In Australia, less-nitrogen demanding Andropogonoid and/or Chloroid genera often replace the more-nitrogen demanding Panicoid genera (Johnson & Tothill, 1985). Adding legumes (such as *Stylosanthes* or *Desmanthus* spp.) to grass pastures delays this replacement by adding nitrogen to the system. Nitrogen-limiting conditions often increase legume establishment. In deep calcareous soils, with high phosphorus, introducing *Leucaena* delayed the decline of pasture grasses (Johnson & Tothill, 1985).

Repair or replace biotic processes in the soil

Soils are complex systems with numerous ecological interactions. In addition to storing nutrients and water, they are the matrix for biological

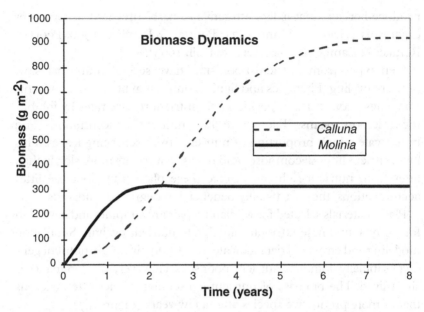

Figure 3.7. Simulated biomass dynamics of a low-productive, nutrient-conserving species (*Calluna vulgaris*) and a high-productive species with high nutrient loss rates (*Molinia caerulea*). This suggests that selecting plant materials for rapid growth and high forage quality may be a mistake on inherently infertile sites since the biomass of nutrient-conserving species will dominate nutrient-poor sites within a few years. From Aerts & van der Peijl (1993). Used with the permission of Munksgaard International Publishers Ltd. (*OIKOS*).

processes that control nutrient cycling (Perry *et al.*, 1989). Soil organisms play essential roles in the carbon cycle and in regulating the availability and cycling of other nutrients (Lee & Prankhurst, 1992). Soil biological properties are degraded by reduced organic matter, reduced biological activity, reduced diversity of soil flora and fauna, and unfavorable changes in biological processes (Lal *et al.*, 1989). Diminished biotic activity adversely affects nutrient cycling, soil physical properties and makes soils less hospitable for plant growth.

Soil structure not only influences hydrologic processes, it affects soil biotic diversity by limiting the movement of larger organisms (Elliot *et al.*, 1980). Smaller organisms, such as protozoans and nematodes, move through very small soil pores on films of water. Flagellates and small amoebae are abundant because they can occupy pore spaces down to 8 μm diameter (Bamforth, 1988). Microarthropod movements are limited because they cannot pass through pore diameters smaller than their bodies (Whitford, 1996). Smaller pores provide refugia for the smaller microbes from larger microarthropod predators. Therefore, the distribution of pore sizes influences the relative abundance and species composition of the microarthropod fauna. This in turn, affects the relative composition of microbial organisms that drive decomposition and mineralization processes (Whitford, 1996).

Disrupting essential ecological linkages within an ecosystem makes that ecosystem more fragile and susceptible to threshold changes (DeAngelis, Post & Travis, 1986). As the physical and chemical condition of a soil deteriorates, soil organisms become less active and perform fewer functions. Since these mutual interactions contribute to ecological stability, restoring those linkages is an essential step toward the repair of damaged ecosystems (Perry *et al.*, 1989). Ecosystem recovery requires the reestablishment and stabilization of energy sources that drive belowground processes (Perry *et al.*, 1989). Since plants provide the energy that fuels these biological processes, the dispersion and growth of plants control the abundance and distribution of free-living symbiotic soil organisms.

Different elements of the soil biota depend on different energy and nutrient sources. Plants and photoautotrophic microbes obtain their energy from sunlight and chemoautotrophic organisms obtain carbon from carbon dioxide in the atmosphere. Specialized plant root–microbial associations can obtain nitrogen directly from the atmosphere. The remainder of the soil biota obtains both energy and nutrients from the soil. Soil organisms may require certain plant species or guilds for their continued existence and do not appear to switch from living plant substrates (as in the rhizosphere) to dead organic matter within the soil. Thus, without plants, soil organisms decline. The composition of the soil bacterial community differs greatly between the rhizosphere and the soil that was not influenced by plant roots. Populations of mycorrhizal fungi and *Rhizobium* spp. decline rapidly in the absence of hosts (Perry *et al.*, 1989). Reducing energy inputs into the soil also affects the physical characteristics of the soil. Many microorganisms, including mycorrhizal fungi, produce extracellular polysaccharides (ECP) that glue mineral particles together into water-stable aggregates (0.25–1.00 mm in diameter) (Perry *et al.*, 1989) that play an important role in soil structure. Clays or polyphenols must protect ECPs, or soil microbes consume them. Without a continuing contribution of

ECPs, soil structure deteriorates. This deterioration accelerates the decline of other soil organisms. Fungi and many bacterial members of the rhizosphere produce resting stages that are subject to consumption by saprophytes and erosion losses unless protected within soil micropores.

Diversity of the plant and microbial communities is believed to be important in stabilizing the system after disturbance (Perry *et al.*, 1989). These links between plants and soil organisms are most important where resources are very limited. Belowground mutualists influence resource availability, both directly by gathering and concentrating nutrients (e.g., nitrogen fixation), and indirectly by improving soil physical properties. Plant diversity stabilizes belowground mutualists. Certain plants form guilds (associations for mutual benefit) with common belowground mutualists. Conifer seedling growth in Oregon and California was greater in soils previously occupied by certain early successional hardwoods (Borchers & Perry, 1987). Soil organisms shared by both hardwoods and conifers were probably responsible for this growth increase.

Mycorrhizae form mutualistic associations with plant roots. Plants benefit from enhanced growth, nutrient uptake, water uptake, and drought tolerance (Allen, 1989) and the fungi receive carbohydrates (energy) from the plant. The vesicular arbuscular (VA) form is common in semiarid and arid lands (Trappe, 1981). Disturbances such as drought or erosion (Powell, 1980), cultivation, and grazing (Bethlenfalvay & Dakessian, 1984; Wallace, 1987) reduce or eliminate the fungi. Mycorrhizal fungi can form hyphal links between roots of different plant species (Newman, 1988). Many of the fungi associated with VA mycorrhizae have low host-specificity. Hyphal links among roots of different species are common since roots of plants intermingle closely. In many natural communities, seedlings commonly establish near older plants. Seedlings link into a mycelium supported by carbon from nearby mature plants. These seedlings, with hyphal links, have rapid mycorrhizal infection (Fleming, 1983; Fleming, 1984) and greater access to inorganic nutrients (Read, Francis & Finlay, 1985). Seedling linkage into mycelium supported by adjacent plants reduces the dominance of large plants and enhances seedling establishment (Grime, 1987). Five-week-old *Plantago lanceolata* seedlings near large mycorrhizal plants were more heavily infected by mycorrhizae than isolated seedlings (Eissenstat & Newman, 1990).

Studies on highly disturbed sites found that mycorrhizae are frequently absent during the early stages of succession. Many of the plants character-istic of these stages are either nonmycorrhizal or facultatively mycorrhizal species (Reeves *et al.*, 1979; Janos, 1980). Disturbed sites on arid Wyoming minelands were initially invaded by opportunistic, ruderal species that did not require mycorrhizal infections (Miller, 1987). While these ruderal species may become infected, most remain nonmycorrhizal. This precip-itates a progressive decline in the number of viable mycorrhizal propagules within the soil that drives the site toward a nonmycorrhizal community that is unlike intact, native communities. However, the inclusion of shrubs initiated a set of processes that began reversing these processes. The shrubs caught windblown propagules: fine soil particles, organic matter, and snow. This small-scale, shrub-enhanced site modification favored the development of mycorrhizal species, principally grasses. These grasses then facilitated the development of more mycorrhizae in the soil, which aided the establishment of more diverse plant communities.

Mycorrhizal fungi are most beneficial where soil resources, water, and nutrients are limiting or where growing seasons are short and plants must exploit resources rapidly (Perry & Amaranthus, 1990). Easily decomposed organic materials, fertilization, and irrigation reduce infec-tion rates of mycorrhizal fungi and rhizobia (Whitford, 1988). Mycorrhizal propagules are absent from many degraded sites and require reintroduction. Under what conditions might mycorrhizal fungi aid repair activities? How is the infection accomplished? Whereas ectomycorrhizal fungi are used to inoculate trees, there are no large-scale methods for inoculation with VA mycorrhizal fungi (Allen, 1989). The most effective strategy is to inoculate container stock in the nursery (St. John, 1990). Seedlings are also inoculated with whole soil from established populations, pieces of root containing mycorrhizal hyphae, or pure cultures of desired organisms (Perry & Amaranthus, 1990).

Add organic materials

Amending soils with organic materials improves the physical, chemical, and biotic properties of that soil. In one study, adding 1% to 6% organic matter (w/w) in the early stages of decomposition (less than one year) to a soil increased aggregate size and decreased erodibility (Chepil, 1955). However, as those initial organic materials broke down over the next four years, they lost their cementing properties. Reduced aggregate stability in the absence of continuing organic inputs causes part of the additional erosion that often occurs during prolonged droughts. Aggregate stability declines in the absence of continuing organic inputs.

Soil physical characteristics are improved with organic materials. Organic materials confer greater resistance to raindrop impact, and reduce wind, and water flow rates. The effect of mulch on erosion on croplands was combined and expressed as:

$$E = Ae^{b \cdot RC} \qquad (3.4)$$

where E is erosion, A and b are constants, and RC is the percentage residue cover (Laflen & Colvin, 1981). The value of b is derived from the relationship between the mulch factor and the crop residue. This equation was used to determine a 'mulch factor' (M_f), by dividing its intercept, A, to give:

$$M_f = \frac{e^{b \cdot RC}}{A} \qquad (3.5)$$

This mulch factor considers the interactions between mulch, soil, and/or slope conditions (Laflen & Colvin, 1981). Organic amendments improve chemical features of soils by adding nutrients, adjusting pH, increasing nutrient conservation, and reducing temperatures that accelerate organic matter losses. Improving the physical and chemical attributes of the soil improves the biological diversity and activity of that soil. The organic matter is also an essential energy source for the soil organisms that perform a number of essential ecological functions.

Plants increase soil organic matter by adding organic matter to the surface, root exudates, and dead root mass. Vegetative and litter cover are increased with management that increases plant production and allows that additional production to be added to the soil. Artificial reseeding and transplanting plant materials increases vegetative cover and may stimulate soil–vegetation development processes. In some situations, it may be feasible to incorporate organic matter into the soil, but this is seldom possible on a large scale.

Although the benefits of organic matter are well known, it is impractical to apply it to large areas. Thus, the application of organic materials to wildlands is usually restricted to relatively small, high-priority sites. There are a great number of potential organic amendments and each provides different benefits (Table 3.4). Selecting a specific organic

Table 3.4. *Comparison of alternative organic matter (OM) sources used to improve degraded soils*

Organic material	Types	Characteristics	Role
Plant residues	straw leaves olive oil cake wood chips bark and sawdust cotton gin trash paper mill sludge	variable OM; stability	stimulate biota; improve physical properties; improve nutrient and water retention
Manure	beef or dairy cattle poultry swine horse	high in degradable OM; nutrient source	stimulate biota; add nutrients; increase nutrient and water retention
Sewage sludge	anaerobic aerobic lime-stabilized waste-activated	nutrient source; lime source if lime-stabilized	stimulate biota; add nutrients; increase nutrient and water retention; increase pH if lime-stabilized
Composts	manure sewage sludge leaf garbage mushroom	high in stable OM; nutrient source	stimulate biota; improve physical properties; increase nutrient and water retention; add nutrients
Peat	sphagnum muck	high in stable OM	stimulate biota; improve physical properties; increase nutrient and water retention

Source: Adapted and modified from Logan (1992). Used with the permission of Springer-Verlag New York, Inc.

amendment depends on its local availability, transportation costs, application costs, and local regulations (Logan, 1992). Inexpensive organic materials provide an important opportunity that may greatly facilitate repair efforts. A few examples illustrate some of the possibilities.

Trees planted on minespoils treated with sewage sludge grow faster than trees on fertilized soils (Berry, 1985). However, the benefits of sewage sludge are short-lived, since it is rapidly decomposed. Digested papermill sludge increased slope stabilization on abandoned mineland because of its fibrous nature (Hoitinek, Watson & Sutton, 1982). Papermill sludge also contains high amounts of free $CaCO_3$ that provide additional benefits when applied to acid soils. Adding 150–300 t papermill sludge ha^{-1} to spoils with a pH of 3.4 maintained the pH at 7.6 for at least three years (Watson & Hoitinek, 1985). Olive oil cake is an industrial by-product obtained from olive oil processing factories in the Mediterranean region. Large amounts of olive oil cake are available at little expense. The water-holding capacity of sandy soils was increased after olive oil cake applications (El Asswad, Said & Mornag, 1992). Thus, where practical, olive oil cake will increase water-holding capacities before planting wildland species. Materials with more stable organic matter (compost or peat) are more desirable for amending degraded soils since they provide longer-lasting benefits (Logan, 1992).

Decomposition is an essential nutrient cycling process (Whitford *et al.*, 1989) that is regulated by water and organic matter availability (Steinberger *et al.*, 1984) and soil biotic diversity (Santos, Phillips & Whitford, 1981; Santos & Whitford, 1981; Elkins, Steinberger & Whitford, 1982; Parker *et al.*, 1984). Decomposition potentials of severely disturbed soils may not recover for many years (Harris, Bentham & Birch, 1991). Respiration-to-biomass ratios (soil metabolic quotient) in mined soils in the Rhineland, Germany, had not stabilized 50 years after mining (Insam & Domsch, 1988), although they improved with each successional change (Insam & Haselwandter, 1989). This possible relationship between the metabolic quotient and vegetative development suggests the potential of influencing the speed, direction, and stability of wildland repair by manipulating the microbial community.

The quality (C:N ratio) of organic materials affects the recovery, persistence, and stability of the soil biota. The diminished biotic diversity and activity of degraded soils (Fresquez, Aldon & Lindermann,

1987; Mott & Zuberer, 1991), reduces the enzymatic capability of the soil microflora, and thus hinders nutrient cycling and organic decomposition. Readily available substrates (following disturbance or fertilization) favor ruderal bacteria. These ruderal species (zymogenous microbes) are r-selected organisms (Andrews & Harris, 1986) and typically dominate following disturbance, but are less abundant under the stable conditions of mature communities. In contrast, autochthonous microbes metabolize difficult-to-degrade organic matter, have slow growth rates, high affinities for growth limiting substrates, and high starvation survival abilities (Andrews & Harris, 1986). The actinomycete bacteria are typical decomposer organisms. They degrade relatively recalcitrant polymeric organic compounds including cellulose, hemicellulose, and lignin. They typically occur in later successional stages and are particularly important in arid ecosystems (Alexander, 1977).

Adding less readily decomposed organic materials, such as bark and wood chips, can accelerate development of more complete soil processes. As an example, in arid ecosystems bark and wood chip amendments contribute to a stable belowground biota that facilitates a more sustainable aboveground flora (Whitford *et al.*, 1989). Bark and wood chips produce a low, but continuous source of organic matter (Whitford, 1988) that is more likely to persist until perennial root systems begin to supply organic matter. The relationship between decomposition rate and the diversity and density of soil microfauna was found in several studies (Santos *et al.*, 1981; Santos & Whitford, 1981; Elkins *et al.*, 1982; Parker *et al.*, 1984). Severely depleted soils treated with readily decomposed organic materials developed soil biota and processes similar to less damaged soils, but the benefits were short-lived (Whitford, 1988). Unlike cultivated soils where nitrogen immobilization by high carbon/nitrogen ratio materials is undesirable, recalcitrant organic materials may be desirable in arid environments (Whitford *et al.*, 1989).

Other hydrologic problems

Dysfunctional hydrologic processes not only reduce soil water for plant growth; they can lead to salinization or gully erosion. Dryland salinization and gully erosion may require landscape-scale repair approaches.

Compacted soils cause severe hydrologic disruptions that prevent autogenic repair processes from working. Severe compaction requires mechanical techniques since it completely disrupts hydraulic processes.

Dryland salinization

Repairing sites affected by dryland salinity requires strategies that lower the water table or reduce recharge rates within the affected watershed. Both simulation modeling studies (Pavelic, Narayan & Dillon, 1997) and actual field applications (Schofield, 1992) confirm the effectiveness of this approach. Modeling studies of a southern Australian watershed indicate dryland salinization is controlled by reducing groundwater recharge over substantial areas (Pavelic *et al.*, 1997). However, small-scale efforts (<100 ha) have no significant impact on groundwater levels. In Western Australia, reforesting 5–10% of a pastureland with a high saline water table reduced the groundwater level 100–200% (Schofield, 1992). Reforesting 25% of the area lowered the water table by approximately 800%.

Planning afforestation to repair dryland salinization requires an understanding of the interception and transpiration potential of alternative species. In Australia, the preferred option for control of dryland salinity is widespread planting of trees and shrubs to lower the water table and thereby reduce salinization (Schofield, 1992). Lowering the water table requires properly selected tree species. Proper trees are adapted to site conditions, transpire enough water to lower the water table, and provide additional products for the landowner. Using trees to lower the saline water table requires that annual evapotranspiration plus streamflow from the land equal or exceed the rainfall and inflow of water from other sources.

Gully erosion

Massive erosion, and the scouring effect of peak flows, hamper plant establishment in channels, slopes, and edges. Watersheds with little vegetation and bare soils create gully problems that increase with slope. Ultimately, gully stabilization requires healthy hydrologic processes on the watershed above the gully. This requires halting damaging activities (abusive grazing, deforestation, inappropriate cultivation, or other soil disturbances) (Duffy & McClurkin, 1967; Heede, 1976; Prajapati & Bhushan, 1993; Morgan & Rickson, 1995c). Then damaged hydrologic processes, leading to excessive velocity and volume of water reaching the gully head, are repaired on the watershed. Each of the soil surface treatments reduces the potential of wind, interrill, and rill erosion. They will also slow the formation and development of gullies by reducing the flow rate of water. Although watersheds with healthy soil surface conditions have a greatly reduced incidence of gully formation, existing gullies may continue their expansion into otherwise stable areas. Engineered structures, biological measures, or some combination of the two, can repair gullies.

Several steps are necessary to repair severe gully erosion (Lal, 1992). However, we must first reduce the causes of damage and address problems on the watershed above the gully. This is achieved by protecting the gully head from livestock and reducing the velocity and volume of water by increasing vegetation in the watershed above the gully or by diverting water away from it. Water should not be diverted through unstable, poorly protected routes. Where eroding faces and beds of gullies are very active, some reshaping is necessary before planting vegetation. Downcutting of the gully bottom and headcutting into the headwater area must be controlled (Heede, 1976).

Engineered structures include diversion channels, drop structures, gabiens, and chutes. They are effective and expensive (in labor, materials, or money) and are not self-maintaining. The benefit/cost ratio of gully control structures is not suitable for most wildland economies, especially in arid and semiarid ecosystems (Lal, 1992). Gully control structures may be either temporary or more permanent structures (Hudson, 1995). Temporary structures provide protection until plants can become established or trap soil where none exists. Temporary structures to trap sediment can be porous and made of netting, brush, and logs or wire cages filled with loose rock. Permanent structures that do not rely on vegetation should only be used as a last resort (Hudson, 1995). They include silt trap dams, drop structures, and gabion structures. Silt trap dams hold large amounts of sediment, so vegetation develops quickly.

Drop structures stabilize the gully head with cement, masonry, or brick. They allow the water to pass harmlessly over them and dissipate the runoff energy. The size of the inlet determines the flow capacity of the structure. Although drop structures seldom fail, they can wash out on the sides or be undermined from below. Gabion structures are heavy-duty wire cages filled with rocks. The primary advantage of gabion structures is the flexibility that allows them to settle and shift without a loss of strength. Engineered structures are most effective when supported with vegetation. Additional design criteria for engineered gully control structures are necessary before planning extensive gully repair efforts (Duffy & McClurkin, 1967; Heede, 1976; Lal, 1992; Prajapati & Bhushan, 1993; Hudson, 1995; Morgan & Rickson, 1995c; Morgan & Rickson, 1995b).

Biological measures depend on establishing vegetative cover (grasses, shrubs, and/or trees) to stabilize bed slopes, improve soil structure, enhance infiltration, and decrease the rate and amount of runoff. Biological gully control measures are less expensive and should be self-maintaining, but are inadequate for the most severe situations. Vegetation alone will seldom stabilize headcuts because of the concentrated flow forces at that point. Plants are most useful where they can control downcutting and grow without engineered structures. The most effective gully vegetation is very thick with deep, dense root systems (Heede, 1976; Morgan & Rickson, 1995b). Flexible plants lie down under flow forces without reducing flow velocities that endanger gully banks and widen the gully, despite channel bottom protection. Large trees that restrict and divert the flow against the bank may move the flow out of the gully and create new channels.

Plants stabilize gullies by (1) reducing the velocity of water flows, (2) dissipating energy that would otherwise be used to detach soil particles, and (3) developing root systems that provide mechanical protection and soil/root cohesion (Morgan & Rickson, 1995c). Reducing water velocity in a channel will drastically reduce the transporting capacity of the water flows. This increases sediment deposition in the channel bottom. While this is often desirable, it can raise the bed level and increase the risk of the water rising out of the channel.

Compacted soils

Water movement through compacted soil layers is greatly restricted and may occur in surface and/or subsurface layers of the soil.

Compacted soils have poor aeration and restrict the physical movement of larger soil organisms. This creates soils with less available water, oxygen limitations, and disrupted nutrient dynamics. These problems not only occur where mechanical equipment has operated, but are also caused by livestock (Stephenson & Veigel, 1987; Lal, 1996). Compaction problems are common on old mine sites (Brown, Johnston & Johnson, 1978; Davies *et al.*, 1992; Ashby, 1997), abandoned roads (Brown *et al.*, 1978; Berry, 1985; Cotts, Redente & Schiller, 1991; Luce, 1997), following timber harvest (Berry, 1985; Guariguata & Dupuy, 1997; Whitman, Brokaw & Hagan, 1997), on old oil-field sites (Bishop & Chapin, 1989b; Chambers, 1989; Whisenant & Hartmann, 1997), and following cultivation (Lal, 1996; Bell *et al.*, 1997). Infiltration is essentially zero in severely compacted soils, making it very difficult to establish plants. Compacted soils usually require deep plowing or deep ripping before plants can establish (Berry, 1985; Ashby, 1997; Bell *et al.*, 1997; Luce, 1997; Whisenant & Hartmann, 1997). However, any treatment that holds water on the soil surface will increase infiltration amounts somewhat. Some compacted soils improve naturally, but very slowly through the actions of freezing, thawing, root penetration, and shrink-swell actions. Chapter 6 describes seedbed preparation treatments for compacted soils.

4

Directing vegetation change

Introduction

Having developed preliminary repair objectives; assessed hydrologic, nutrient cycling, and energy capture processes; and designed alternative strategies for repairing damaged primary processes, we need strategies for directing vegetative change. Now it is necessary to devise management strategies that continue developing the vegetation to: conserve soil, nutrient and organic resources; return fully functional hydrologic, nutrient cycling and energy capture processes; and create self-repairing landscapes that provide the goods and services necessary for ecologic and socioeconomic sustainability.

Will the site recover within an acceptable time frame in the absence of active repair efforts? If so, will it provide the desired combination of goods and services? Improving the management of ecosystem consumption (i.e., livestock, timber harvest, fodder harvest) is usually the best strategy for relatively intact sites. Additional degradation makes it necessary to actively manipulate the existing vegetation by reducing some species (with fire, herbicides, mechanical, or biological control methods) and/or adding others (with seeding or transplanting seedlings). Denuded or depauperate sites that can neither stabilize the site nor achieve management objectives require the addition of more plants.

Species performance, site availability, and species availability influence the direction and pace of vegetation change. Each of these causes has a set of contributing processes or conditions and defining factors. This organization of causes, processes and defining factors provide a convenient format to design repair strategies for specific problems. These management actions range from those with immediate effects (e.g., weed control, plant removal, or seedbed preparation) to those actions with longer-term objectives (e.g., establishing shrubs on degraded sites to collect wind-blown seed or attract birds that import seed).

Understanding vegetation change

Vegetation change is 'any dynamic vegetation pattern where dominant populations of one or more species on a site are being replaced by new populations of the same or different species' (Burrows, 1991). Thus, succession is a subset of vegetation change since it requires species replacement and vegetation change does not. Ecosystems with extreme environmental conditions or otherwise difficult plant growth conditions do not always undergo sequential replacement of plant species. Where the initial colonizers of a damaged site persist indefinitely, without being replaced, subsequent community development is vegetation change rather than succession (Burrows, 1991). Damaged wildlands require vegetative development and the addition of organic materials to increase biotic controls over limiting resources. This 'vegetative development' might be either succession or vegetation change. Either way, we seek to initiate and direct autogenic processes leading toward self-repairing wildlands that provide necessary goods and services. Our ability to accomplish this is limited by our understanding of how plant communities develop following disturbance.

Early successional models often assumed that a single climax community existed for each site, and that successional changes on improving landscapes simply reversed the changes that occurred during degradation (Clements, 1916; Clements, 1936). This implied a climax community could be achieved simply by improving management (e.g., grazing management or timber harvest), although experience has indicated recovery often requires significant management intervention (Friedel, 1991). Many of the difficulties encountered when repairing damaged wildlands are the consequence of viewing succession as an orderly, sequential change toward a predetermined climax vegetation. For example, a widely used guideline suggests that if at least 15% of the existing species are desirable, improved management is sufficient to repair the damage (Vallentine, 1989). Although based on a concept of succession that is increasingly discredited in ecological circles (McCook, 1994), it remains a dominant force and an underlying assumption of many management programs. More contemporary theories suggest this deterministic view of succession is an exception rather than the rule (Friedel, 1991; Laycock, 1991).

A contemporary view of succession and vegetation change has evolved that now reflects (1) the importance of both process and context; (2) the inherent uncertainty in biotic and abiotic events and the role of infrequent and rare events; (3) the temporal and spatial variability inherent in most wildlands; (4) the realization that succession is the cumulative effect of many plant-to-plant events, rather than a single operative mechanism; (5) the importance of multiple, relatively stable vegetative states and threshold events. Each of these factors makes it problematic to set specific vegetation goals for repair efforts.

Process and context

Processes operating outside the site may regulate vegetation change on a particular site (Pickett *et al.*, 1992). Process 'incorporates movement and interaction among organisms, the transformation of energy and material, and vegetation change, changes in patchiness, or responses to environmental change' (Pickett & Parker, 1994). Processes arising outside the boundaries of an ecological system help regulate its functioning. These interactions, with the surrounding landscape, define the context of that ecological system. This new view of ecological systems suggests the importance of both process and context in the repair of wildland ecosystems. Wildland repair programs must also consider the socioeconomic context, which includes economic, esthetic, religious, policy, governmental and perhaps even subsistence issues. Chapters 2 and 8 contain additional information on the importance of ecological context. Chapter 8 also discusses the socioeconomic context.

Uncertainty, rare and infrequent events

Rare and unusual events play a more important role in shaping vegetation than we previously understood. Establishment and death of some species is episodic, requiring unusual climatic events. Successful reproduction or mortality of some species only occurs following rare events, but those events shape the landscape for a long time. This is most obvious with long-lived woody species, but similar situations also occur with herbaceous species. In arid ecosystems, unusual precipitation events trigger widespread germination and short-term establishment. The episodic establishment of Mitchell grass (*Astrebla pactinata*) in Australia is believed to coincide with certain phases of the El Niño Southern oscillation that produces fall precipitation (Austin & Williams, 1988). Once established, Mitchell grass plants persist for a long time and dominate large areas, but its very presence depends on rare events.

Numerous historical and environmental circumstances combine in unique ways to insure that a universal, general cause for succession will not be found (McCook, 1994). Unique combinations of abiotic limitations and biotic interactions act stochastically to change vegetation. Vegetation development alters the environment through gradual accumulations of soil organic matter, enhanced nutrient pools, and altered microenvironmental conditions. These changes produce environmental gradients that sort species based on life history and resource allocation strategies.

Temporal and spatial variability

Contemporary successional models incorporate spatial variation, lag effects, thresholds, event-driven changes, and dynamic rather than static equilibria (Walker, 1993; Wyant *et al.*, 1995). The concept of stable ecosystems has been transferred to larger temporal and spatial scales (Sprugel, 1991). Stability at small spatial scales is probably an inherently short-term phenomenon, but stability becomes more common at larger spatial scales and over longer periods (DeAngelis & Waterhouse, 1987). Ecosystems with a high disturbance frequency can be in equilibrium if the creation of new patches is balanced by the maturation of old patches (Sprugel, 1991). While small patches may appear as constant change, they may represent equilibrium between disturbance and succession at the landscape scale. Newer successional models recognize that equilibria only exists at larger spatial scales (DeAngelis & Waterhouse, 1987; Wyant *et al.*, 1995).

Since natural systems contain temporal and spatial variation that varies in complex ways at several scales (White & Walker, 1997), we cannot fully understand all the changes that might occur following a repair program. Nor should we rely on reference sites for unquestioned goals.

Multiple mechanisms of change

The cumulative outcome of numerous plant-to-plant interactions determines successional outcomes, rather than single mechanisms (Figure 4.1). Although single-mechanism successional models do not adequately portray the full range of succession sequences, a three-pathway successional model is useful (Connell & Slatyer, 1977). Facilitation, the first pathway, occurs if the early occupants change the environment enough to allow new species to establish. It is also widely viewed as 'the process in which two individual plants or two populations of plants interact in such a way that at least one exerts a positive effect on the other' (Vandermeer, 1989). This is a very different view of facilitation since it does not involve replacement of the original species, but it is extremely useful when incorporated into wildland repair programs. Facilitation without species replacement is probably a more common phenomenon than facilitation that results in species replacement. This view of facilitation (without replacement of species) is fundamental to the agroforestry and intercropping disciplines, which have developed theory and management strategies to encourage positive interactions among plant species (Vandermeer, 1989; MacDicken & Vergara, 1990; Nair, 1993). These disciplines provide an important source of information for directing facilitation in wildland repair situations.

The second pathway, tolerance, describes the situation in which later species are unaffected by earlier species; they establish and grow to maturity in the presence of previous species because they can grow at lower levels of resource availability or use different sources of limiting resources. With inhibition, the third pathway, early species prevent the growth and maturation of subsequent species. Each of these three successional pathways provides numerous opportunities for influencing the direction and pace of vegetation change (Figure 4.1).

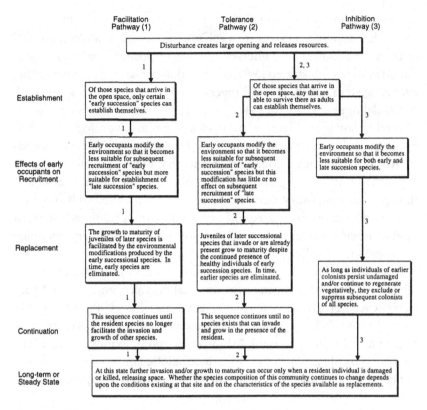

Figure 4.1. Three alternative mechanisms producing the sequence of species change during plant succession. These pathways are most effectively thought of as interactions between individual plants – with the cumulative effect producing species changes. A given site may have all three of these pathways operating among different species at the same time. From Connell & Slatyer, (1977). Used with the permission of The University of Chicago Press (*The American Naturalist*).

Multiple stable-states and transition thresholds

Where succession in a particular ecosystem has the potential for more than one stable vegetative states, separated by transition thresholds, management changes alone do not necessarily lead to recovery. Transition thresholds, controlled by abiotic limitations or biotic interactions, limit our ability to direct vegetative development (illustrated in Figure 1.1). Thus, redirecting succession often requires active management intervention (Friedel, 1991).

TRANSITIONS CONTROLLED BY ABIOTIC LIMITATIONS

Degraded soil surfaces create problems with water infiltration and nutrient retention. These interactions trigger positive feedback degradation systems (illustrated in Figure 2.6) that contribute to catastrophic events (system collapse) and irreversible vegetation changes (Rietkerk & Vandekoppel, 1997; van de Koppel, Rietkerk & Weissing, 1997). Severely damaged primary processes such as hydrology, nutrient cycling, and/or energy capture often create transition thresholds controlled by abiotic limitations. Treatments that focus on roughening the soil surface or creating aboveground obstructions provide essential, though only temporary, benefits. They are implemented to halt degradation (e.g., erosion and runoff) and improve conditions for plant establishment and growth until autogenic processes can dominate (Whisenant, 1995; Whisenant *et al.*, 1995).

Degraded wildlands experience greater temperature extremes and higher wind speeds. Those environmental changes, and reduced soil-water levels, create difficult abiotic environments (Uhl, 1988; Lugo, 1992; Brown & Lugo, 1994; Guariguata, Rheingans & Montagnini, 1995; Fimbel & Fimbel, 1996; Ashton *et al.*, 1997). In most wildland situations, the only practical way to reduce abiotic limitations is by initiating plant growth that will contribute to autogenic development.

TRANSITIONS CONTROLLED BY BIOTIC INTERACTIONS

Transition thresholds controlled by biotic interactions are caused by inhibition, limited propagule availability, damaging management practices, or more commonly, a combination of several factors. For example, in the subtropical, semiarid savannas of south Texas, the conversion of semiarid grasslands to woodlands is essentially irreversible (Archer, 1989). Abusive grazing practices damaged the original grassland-dominated system and altered the composition. These changes reduced the herbaceous productivity and decreased the amount of fine fuel. Less fine fuel (herbaceous vegetation) decreased the frequency and intensity of fire, and disrupted the natural fire regime (and seed transport by livestock). These changes increase the establishment of woody plants until the system crosses a threshold toward a shrub-driven system. This initiates a positive feedback system in which shrubs continue to increase, herbaceous production decreases, the carrying capacity for grazers is decreased, and the grazing pressure in the remaining interstitial zones become greater. Once in this shrub-driven state, the soils, seedbank, and vegetative regenerative potentials are altered and the site will not revert to grassland or savanna, even if it is no longer grazed by livestock (Archer, 1989).

Traditional succession models cannot predict changes in some ecosystems because of irreversible transitions and alternate stable states. In the California Mediterranean grasslands, fire maintained open grasslands (George, Brown & Clawson, 1992). With increasing grazing pressure, herbaceous vegetation decreased, woody seedlings increased, and water moved deeper into the soil profile. This causes positive feedbacks that accelerate the recruitment and growth of woody plants. Thus, the spatial distribution of nutrients and soil organic matter changed from the relatively uniform distribution of grasslands to the patchy distribution of woodlands. Trees increase the proportion of the total rainfall that gets into the subsoil by creating preferential flow paths. This creates an unstable threshold separating two relatively stable vegetative states. One is a woodland with little grass and the other contains a mixture of woody and herbaceous vegetation (George *et al.*, 1992).

Setting goals

Directing vegetation change on damaged wildlands requires clear and achievable objectives. Since we seldom understand the composition, structure, function, or dynamics of historic ecosystems (Sprugel, 1991) and they are very difficult to determine (Miller, 1987), it is often unrealistic to measure the success of a repair program against historic conditions. Repair goals should reflect the understanding that wildlands are changing, dynamic systems rather than static and unchanging (Johnson & Mayeux, 1992; Pickett & Parker, 1994; Wyant *et al.*, 1995). While it may be impractical to set specific reference conditions as precise goals, using reference ecosystems from similar landform/soil/climatic conditions to guide planning efforts is useful (Hobbs & Norton, 1996). Unless site degradation has been excessive, reference ecosystems provide a first approximation of the type of vegetation best suited to a particular landform, soil type, and climatic conditions. Hobbs and Norton (1996) listed several potential attributes for consideration:

1. Composition: existing species and their relative abundance.
2. Structure: vertical arrangement of vegetation (living and dead).

3. Pattern: horizontal arrangement of vegetation (living and dead).
4. Heterogeneity: a complex variable made up of vegetative components 1–3, soil characteristics, and litter distribution.
5. Function: performance of essential ecological processes (energy capture, hydrology, nutrient cycling).
6. Vegetation dynamics and resilience: successional processes, recovery from disturbance.

After ranking the relative importance of various attributes, we must decide how closely the repair site should resemble the reference ecosystem.

Directing change

The diversity of ecosystems and management objectives insures no 'cookbook' approach can possibly have universal application. Wildlands in different stages of degradation require initial management actions that focus on different processes. Chapter 3 described strategies for severely degraded sites that require immediate repair of primary processes, primarily at the soil surface. These soil surface modifications increase perennial vegetation that continues to improve hydrologic and nutrient cycling conditions. As it develops, vegetation reduces abiotic limitations by improving soil and microenvironmental conditions. Directing autogenic processes toward our landuse goals requires an understanding of the processes driving succession and vegetation change and where necessary. Three basic successional causes may be manipulated to direct succession: (1) differential species performance; (2) differential site availability; and (3) differential species availability (Luken, 1990). These causes, processes, and defining factors of succession suggest a framework for considering concepts relevant to the design of specific repair strategies (Figure 4.2).

Differential species performance

Differential species performance occurs when a species, or group of species, outcompete other species. The relative performance of some species over other species is influenced by: resource availability, eco-physiology, life-history strategy, environmental stress, competition, allelopathy, disease, herbivory and predation (Rosenberg & Freedman, 1984; Pickett *et al.*, 1987b). Each of these processes can be manipulated with traditional methods (e.g., grazing management, plowing, or weed control), ecological methods (e.g., by encouraging autogenic development, inhibition, seed vectors, or seed predators), or some combination of each. Our ability to direct vegetative change by differentially manipulating species performance is enhanced with a good conceptual understanding of individual plant-to-plant inter-actions, resource availability, life history, ecological strategies, and herbivory.

PLANT-TO-PLANT INTERACTIONS

Facilitation, tolerance, and inhibition are comparative statements of plant-to-plant interactions, not mechanisms of entire successions (McCook, 1994). All three occur simultaneously in many communities. Although each of these concepts provides information for wild-land repair, facilitation is most useful. As plants and their associated organisms grow, they change the environment through a process called 'reaction' (Clements, 1916; Clements, 1936). Reaction is the effect of plants on the physical environment and the subsequent positive feed-back to the vegetation. Facilitation, positive interactions, autogenic influences, and ecosystem engineering describe similar processes. This environmental modification by plants occurs through both passive and active mechanisms. Plants passively affect their immediate environments with their physical structure by shading the soil and altering wind movements (Figure 4.3). This reduces wind speed, lowers the extremes of air and soil temperatures, and increases relative humidity. Plant structures trap wind-blown soil, nutrients, and propagules of microorganisms and other plants. Metabolic processes actively change the environment by altering temperature, humidity and the physical and chemical properties of soils. Plants gradually increase soil organic carbon and improve the water and nutrient holding capacities of the soil. The capacity of plants to modify their environment is roughly proportional to vegetation biomass, stature, and the rate of metabolic activity (Roberts, 1987). Thus, sparse desert vegetation is less able to alter its environment than forest vegetation. However, the lesser

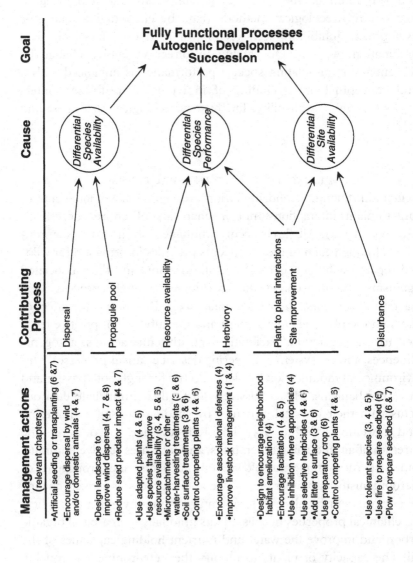

Management actions
(relevant chapters)

**Contributing
Process**

Cause

Goal

**Fully Functional Processes
Autogenic Development
Succession**

*Differential
Species
Availability*

*Differential
Species
Performance*

*Differential
Site
Availability*

Dispersal

Propagule pool

Resource availability

Herbivory

Plant to plant interactions

Site improvement

Disturbance

•Artificial seeding or transplanting (6 &7)
•Encourage dispersal by wild
and/or domestic animals (4 & 7)

•Design landscape to
improve wind dispersal (4, 7 & 8)
•Reduce seed predator impact (4 & 7)

•Use adapted plants (4 & 5)
•Use species that improve
resource availability (3, 4, 5 & 3)
•Microcatchments or other
water-harvesting treatments (3 & 6)
•Soil surface treatments (3 & 6)
•Control competing plants (4 & 6)

•Encourage associational defenses (4)
•Improve livestock management (1 & 4)

•Design to encourage neighborhood
habitat amelioration (4)
•Encourage facilitation (4 & 5)
•Use inhibition where appropriate (4)

•Use selective herbicides (4 & 6)
•Add litter to surface (3 & 6)
•Use preparatory crop (6)
•Control competing plants (4 & 6)

•Use tolerant species (3, 4 & 5)
•Use fire to prepare seedbed (6)
•Plow to prepare seedbed (6 & 7)

Figure 4.2. Diagrammatic representation of how management actions might be used to influence vegetative development by focusing on contributing processes and causes of change.

Figure 4.3. Rapidly growing shrubs (*Acacia holoserica*) were transplanted into microcatchments on this previously barren site in Niger. This illustrates the importance of physical modifications (microcatchments) to overcome abiotic limitations and allow shrub establishment. As the shrubs developed they improved hydrologic, nutrient cycling, and microenvironmental conditions (neighborhood habitat amelioration) enough to facilitate the natural establishment of numerous herbaceous species.

plant-induced environmental alterations in arid ecosystems may still have significant biological impacts.

Positive interactions among species occur under two contrasting, but relatively predictable sets of circumstances. The predictability of these positive interactions is important because it allows us to more reliably incorporate them into ecological repair programs. Positive interactions are believed to be most common in communities (1) that developed with high physical stress; and (2) that developed with intense consumer pressure (Bertness & Callaway, 1994). Communities with intermediate levels of physical stresses and consumer pressures should have the fewer positive interactions and more competitive interactions (Figure 4.4). Harsh physical conditions favor neighborhood habitat amelioration (Bertness & Callaway, 1994).

The physical stature of woody plants has strong ameliorating influences on the microenvironment and soil of its immediate surroundings

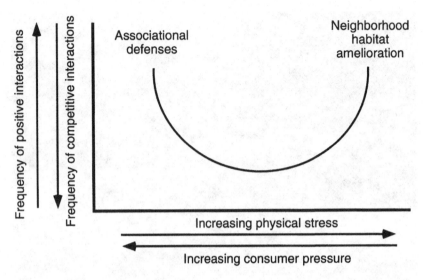

Figure 4.4. Conceptual model of the occurrence of positive interactions in natural communities. Positive interactions are predicted to be rare under mild physical conditions and low consumer pressure. Amelioration of physical stress by neighboring plants is believed to be most common under harsh physical conditions. Associational defenses are believed to be most common under intense consumer pressure. From Bertness & Callaway (1994). Used with the permission of Elsevier Science (*Trends in Ecology and Evolution*).

(Figure 4.5). This relationship is pervasive in the literature of ecology, ecological restoration, and agroforestry. In arid and semiarid ecosystems, shrubs or trees improve microenvironmental conditions by moderating wind and temperature patterns (Allen & MacMahon, 1985; Farrell, 1990; Vetaas, 1992; Whisenant *et al.*, 1995; Rhoades, 1997). Although the woody plants compete with understory plants for light, the benefits of this habitat amelioration often outweigh any negative effects (Holmgren, Scheffer & Huston, 1997). Agronomic crops sheltered by wind barriers tend to grow taller, produce more dry matter, have a larger leaf area index, and larger yields (Vandermeer, 1989). Juvenile pines (*Pinus strobus* and *Pinus resinosa*) occur beneath oak (*Quercus rubra*) canopies at densities over six times that occurring in open areas of Ontario, Canada, but not until the oaks were at least 35 years of age (Kellman & Kading, 1992). This delayed effect suggests the importance of physical stature.

Associational defenses are also important. Where livestock grazing

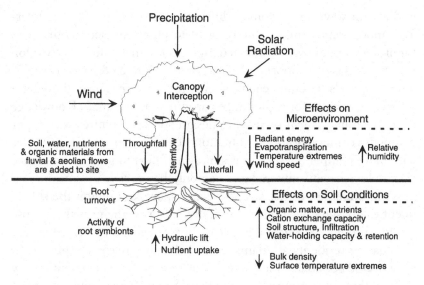

Figure 4.5. Factors influencing microenvironmental and soil conditions under and around an individual shrub or tree. After Farrell (1990). Used with the permission of Springer-Verlag New York, Inc.

has a strong influence, palatable herbaceous species are often restricted to growing under the physical protection of thorny shrubs or cacti. While there may also be some habitat amelioration, protection from grazing (associational defense) appears to be the strongest factor. Intercropping studies provide information on associational defense mechanisms that may also reduce insect damage in wildlands. Three mechanisms are believed to be responsible for this protection (Vandermeer, 1989). First, the disruptive crop hypothesis where a second species disrupts the ability of a pest to efficiently attack its proper host (more common with specialist insects). Second, the trap-crop hypothesis where a second species attacks a pest that would normally be detrimental to the other species (more common with generalist insects). Third, the enemies hypothesis where the species mixture attracts more predators and parasites than monocultures and reduces pests through predation or parasitism. These mechanisms suggest additional benefits of diverse species mixtures.

Species minimize competition and coexist by using different resources, using resources at different times, using the same resources

in different ways, or exploiting different ecological niches. The tolerance model represents a suite of plant-to-plant interactions that lead
toward a community dominated by species that efficiently exploit
different kinds or proportions of resources (Connell & Slatyer, 1977).
Those species tolerate each other's presence because they have
different strategies for exploiting environmental resources. Knowledge
of these strategies can be used to design species mixtures with fewer
competitive interactions. Our attempts at developing species mixtures
with more compatible species is simply an effort to increase tolerance
(i.e., reduce negative interactions) among the species.

With inhibition, earlier species secure space and inhibit the subsequent establishment of other species or suppress the growth of those
already present. Suppressed species invade or grow only when the
dominating residents are damaged or killed and release resources. In
the extreme possibility, the first colonists prevent establishment of new
species, thus preventing subsequent succession. This inhibition of late
arrivals may either be the result of competition (i.e., resource depletion) or allelopathy (Connell & Slatyer, 1977; McCook, 1994).

With the inhibition model, early succession species may be as resistant to invasion as late species, so mature species are most resistant to
damage by fires, storms, natural enemies, etc. (Connell & Slatyer,
1977). When species replacement only occurs after the damage or
death of existing vegetation, replacement ends in the eventual domi-
nation of longer-lived species (Connell & Slatyer, 1977). Long life is
partially the result of having defenses against, or tolerance of, all the
inevitable hazards. Juveniles of late-succession species develop deeper
and more extensive root systems than early successional species,
allowing them to persist through occasional drought periods. Fire and
flood alluvium kill several tree species that invade gaps and suppress
young redwood trees, but the redwoods remain unharmed through
those disturbances (Hollick, 1993). The hard wood and thick bark of
redwoods requires more energy and material, slowing tree growth, but
it probably confers enhanced survival following disturbances such as
fire and flood. In parts of the western United States, shrubs may dominate previously forested sites for more than 100 years after trees are
removed by fire or logging (Radosevich & Holt, 1984). This probably
results from competition that intensifies resource limitation, thus limiting tree growth and slowing succession. In Great Britain, heathland
and moorland communities require bracken fern (*Pteridium aquili-*

num) control before they begin to recover. Even after bracken control, recovery may be inhibited by a severely depleted seedbank of native species if bracken had dominated the site for a long period (> 50 years) (Pakeman & Hay, 1996).

RESOURCE AVAILABILITY

Enduring low resource levels is an active tolerance mechanism, whereas the use of coexisting plants with contrasting life-history strategies is a passive tolerance mechanism. These two strategies overlap since rapid resource use is often correlated with a short life cycle, early maturity and abundant reproductive output (Pickett *et al.*, 1987b; Pickett, Collins & Armesto, 1987a). Plants that depend on high rates of resource use cannot tolerate low resource levels (Grime, 1977). Resource use rates are related to competitive ability, since effective competitors acquire resources before weak competitors (Pickett *et al.*, 1987b; Pickett *et al.*, 1987a). In forest successions, later successional trees typically replace pioneer trees that require more sunlight and are less tolerant of shade. During succession, each group of species is more shade tolerant than previous species. As the forest canopy thickens, shade tolerant species start to dominate. Tolerance to variations in other environmental factors (moisture, temperature, nutrients, grazing, salinity, etc.) may be just as important in other situations (Connell & Slatyer, 1977).

The traits of early plant colonists allow them to take advantage of readily available resources on disturbed sites (Grime, 1977; Chapin, 1980). Pioneer plants grow rapidly as long as resources are readily available. They have rapid leaf turnover and invest little energy into secondary chemical compounds for defense against herbivores (Coley, Bryant & Chapin, 1985; Coley, 1988). As a result they recover rapidly following defoliation, but this advantage decreases with time (Chapin, 1980). Many plants of late succession or very infertile soils have evolved traits that conserve nutrients and reduce herbivory. This is accomplished with reduced growth rates, longer tissue life, and greater metabolic investments into mechanical defenses (e.g., thorns) (Owen-Smith & Cooper, 1987) or secondary metabolites (Coley, 1988). Herbivores often prefer pioneer species to later successional species because of these plant defenses (Davidson, 1993). Consequently, this preference for early successional species is believed to accelerate succession in some ecosystems (Bryant & Chapin, 1986; Walker & Chapin, 1986). In other ecosystems, the pioneer vegetation might be less palatable than the intermediate stages. An extensive literature survey, of studies ranging from African savannas to boreal forests, concluded that the influence of herbivory on successional changes was predictable (Davidson, 1993). The most commonly consumed species occurred in intermediate successional stages, in favorable resource environments, and responded to grazing with rapid, compensatory growth. In this environment, herbivory tends to favor pioneer species and less palatable species.

'In any system in which a critical source has a flow rate through the system, there is always a possibility that one species might act to reduce the rate of flow of the resource out of the system, with the indirect consequence that another species may benefit' (Vandermeer, 1989). This is particularly relevant to nitrogen since its rate of movement through soils can be rapid. Woody plants can reduce nutrient and water losses from disturbed landscapes by capturing wind-blown organic materials, soil particles, nutrients (Virginia, 1986) and microorganisms (Allen, 1988b). Perennial, nitrogen-fixing legumes are believed to be essential components of many ecosystems (Knoop & Walker, 1985; Jenkins *et al.*, 1987; Jarrell & Virginia, 1990), because of their ability to develop symbiotic associations with both rhizobial bacteria and mycorrhizal fungi (Herrera, Salamanca & Barea, 1993). Woody plants redistribute nutrients through root harvesting, alter air turbulence patterns, and influence patterns of deposition, and organic matter (West & Caldwell, 1983). The aerodynamic qualities of woody plants capture blowing vesicular arbuscular mycorrhizae (VAM) and saprophytic fungal spores, and organic matter (Allen, 1988b). Soil nutrients such as available phosphorus (P) may increase down slopes because of particle deposition and wind movement (Allen, 1988b).

Plants have differential access to nutrient pools because of different rooting depth and growing season. Intercropping studies have demonstrated that certain species mixtures take up more soil P than do monocultures of the same crops (Vandermeer, 1989). One of the crops may use nutrients that are unavailable to the other crop. However, those previously unavailable nutrients may become available after cycling through the first crop. Nutrient transfer between species can occur through mycorrhizal connections (Chiarello, Hichman & Mooney,

1982). Thus, since mycorrhizal connections regularly form between species, the ability to transfer otherwise unavailable nutrients among species is facilitation (Vandermeer, 1989).

Although competition for water occurs at some time in virtually all ecosystems, occasionally one species improves the water environment for other species. Soils around individual trees or shrubs commonly have higher soil organic matter, available water, and reduced evaporation, compared with soils found between woody plants. Deeply rooted plants can increase the amount of water in the surface soil through the process of hydraulic lift. Woody plants also trap snow and reduce losses to wind and sublimation, thus improving soil water relations for associated species (West & Caldwell, 1983). Thus, the stature of individual plants, or patches of vegetation, play an important role.

HERBIVORY

Herbivory affects competitive interactions among plant species. Thus, it can determine the outcome of wildland seeding activities. Since seedlings have low nutrient and energy reserves and poorly developed root and shoot systems, it is important to delay grazing by domestic livestock and wildlife until the seedlings are well established. Poor grazing management reduces the diversity, productivity, and persistence of the desired vegetation (Whisenant & Wagstaff, 1991). The influence of aboveground foliage removal is widely appreciated, but belowground grazers may consume more plant material (Stanton, 1988). Nematodes and insect larvae may significantly reduce shoot production and increase plant mortality (Ueckert, 1979; Stanton, Allen & Campion, 1981). Nematodes can decrease plant production by 6–13% in grassland ecosystems (Ingram & Detling, 1984). Following grazing, grasses are more susceptible to parasitic nematodes (Stanton, 1983; Ingram & Detling, 1984). Site preparation techniques that kill existing vegetation may reduce belowground herbivore populations by reducing their food supply (Archer & Pyke, 1991). Interseeding or broadcast seeding techniques that do not kill existing vegetation (and belowground herbivores) may potentially limit seedling establishment.

Differential site availability

The most critical phase of most wildland repair projects is seedling establishment. Most failures occur during seedling establishment. Understanding how to manipulate seedbed environments provides numerous opportunities to increase seeding success. In the short term, the availability of safe sites is the ecological filter that selects for or against a particular species. In the long term, site changes resulting from autogenic development become more important. Safe sites and autogenic development are addressed in this section. Chapters 6 and 7 address the more traditional aspects of seedbed preparation and planting.

SAFE SITES

The majority of seed within a seedbank do not germinate in any particular year. Of the many seeds moving into a seedbank, only a small number produce seedlings and only a fraction of the seedlings become established (Urbanska, 1997). As a rough approximation, we can relate the number of seedlings produced to the number and distribution of safe sites in the seedbed. A safe site is '. . . that zone in which a seed may find itself which provides (a) the stimuli required for breakage of seed dormancy, (b) the conditions required for germination processes to proceed and (c) the resources (oxygen and water) which are consumed in the course of germination. In addition, a "safe site" is one from which specific hazards are absent – such as predators, competitors, toxic soil constituents and pre-emergence pathogens' (Harper, 1977).

Seed size and shape, relative to that of adjacent soil particles, play an important role in determining water availability to the seed. These relationships between soil surface features and seed morphology control seed entrapment and influence seedling establishment patterns, particularly on exposed soils (Chambers, 1995). For example, yarrow (*Achillea millefolium*) has flat seeds that germinate best on an even soil surface. However, species with other seed shapes germinated poorly on a flat soil surface and did much better in 20-mm (0.8 inch) grooves (Oomes & Elberse, 1976).

Plants modify the abundance and characteristics of safe sites within a seedbed. Shading, chemical changes in the soil, and litter accumulations affect the microenvironment and availability of safe sites

(Harper, 1977). They alter the relative occurrence of safe sites for each species, since a safe site for one species may not be a safe site for another species. In forests, the relative shade tolerance of seedlings greatly influences seedling establishment and the composition of the resulting vegetation. Many species only establish under full sunlight, while others are extremely tolerant of shading. Shade is less of a factor in grasslands, but the distribution of leaves and litter has important selective actions in grassland communities. The intentional use of vegetation, as 'nurse plants,' to increase the establishment and growth of planted species is described in more detail in Chapter 6.

The importance of the microenvironment in grassland seedling establishment has been well established (Harper, Williams & Sagar, 1965). Soil surface topography and plant litter are important factors in the establishment of downy brome (Evans &Young, 1984). Litter or rocks increased germination, survival, and growth of adjacent *Aristida longiseta, Bouteloua rigidiseta,* and *Stipa leucotricha* in central Texas (Fowler, 1986). Soil surface features, such as plant litter or rocks (Figure 4.6) function as safe sites by creating more mesic microsites.

Shade produced by large rocks or adult saguaro cacti (*Cereus giganteus*) controlled saguaro seedling recruitment in the Sonoran Desert (Turner *et al.*, 1966; Steenbergh & Lowe, 1969). In the Chihuahuan desert, shrubs made the immediate surroundings more suitable for the natural recruitment of perennial herbaceous species (Whisenant *et al.*, 1995). Palatable vascular plants derive associational benefits from living among less palatable neighbors (Bertness & Callaway, 1994). Thorny shrubs and cacti (*Opuntia*) provide protection for much of the herbaceous production in overgrazed *Acacia* and *Prosopis* savannas. Although some of this benefit involves microenvironmental modification, it also suggests the benefits of protection from large herbivores.

Organic materials on the soil surface are important modifiers of the seedbed environment. Litter increased downy brome seedling establishment in arid sagebrush grasslands by moderating air temperature and water at the surface soil (Evans *et al.*, 1970). Japanese brome (*Bromus japonicus*) density in April was a function of litter accumulations and autumn precipitation, with the importance of litter accumulations increasing as autumn precipitation decreased (Whisenant, 1990). Not all species benefit from surface litter accumulations. Lehman's lovegrass (*Eragrostis lehmanniana*), a warm-season perennial

Figure 4.6. Seed were broadcast over this west Texas (about 400 mm precipitation) slope without any attempt to cover the seed. The cracks and crevices under and around the rocks created numerous safe sites for seedling establishment. The presence of rocks in the soil matrix maintained the macropores and conferred significant protection against compaction, runoff, or reduced infiltration. Broadcast seedings on smooth, bare soil without rocks or seedbed treatments have low success rates.

grass adapted to arid conditions, increased recruitment following fires that removed plant litter (Ruyle, Roundy & Cox, 1988).

Decaying logs may act as safe sites (or nurse logs) for tree seedlings (Christy & Mack, 1984; Scowcroft, 1991; Lusk, 1995; Szewczyk & Szwagrzyk, 1996). In Oregon, (Christy & Mack, 1984) 98% of western hemlock (*Tsuga heterophylla*) juveniles occurred on decaying Douglas fir (*Pseudotsuga menziesii*) logs, although these logs covered only 6% of the total available area. They found the seeds germinated on logs of other trees and on mineral soil, but seedling survival was better on Douglas fir logs. In the Hakulau Forest of Hawaii, 53–70% of native tree regeneration occurred on decaying logs which made up less than 2% of the potential seedbed area (Scowcroft, 1991). Although mineral soil accounted for about 97% of area, only 25–33% of regeneration occurred there.

AUTOGENIC DEVELOPMENT

Environmental modification by the vegetation has profound implications for the repair of damaged wildlands. Since vegetation responds to and strongly modifies its immediate environment (soil and microclimate), we should direct those changes to meet repair objectives. For example, once converted to pastureland, tropical rainforest recovery is inhibited by a lack of seeds, seed predation and a harsh microclimate that kills developing seedlings (Uhl, 1988). Therefore, repairing degraded tropical forests requires strategies that (1) increase natural seed dispersal; (2) reduce the impact of seed predators; and (3) ameliorate harsh microenvironmental conditions. Overcoming these limitations is very costly unless natural processes are stimulated to achieve those objectives. Because interior forest species cannot tolerate existing environmental conditions, it may not be possible to begin with native rainforest species. However, several studies have demonstrated the feasibility of planting adapted tree species (either native or exotic) as a first step toward developing a native forest (Uhl, 1988; Lugo, 1992; Brown & Lugo, 1994; Guariguata et al., 1995; Fimbel & Fimbel, 1996; Ashton et al., 1997).

In Puerto Rico, tree plantations modify soil and microenvironmental conditions enough to facilitate the natural immigration of native species (Lugo, 1992). As the plantation trees grew, they altered microenvironmental conditions, making them more suitable for native forest species. The native species accumulated litter on the forest floor, which increased nutrient retention and reduced erosion. The plantation trees also accelerated the return of native plant species by attracting animals that import seed. Tree plantations in the moist and wet tropics do not remain monocultures (Lugo, 1992), because native trees invade the understory and penetrate the canopy of the exotic species. In the absence of extreme site damage, native forests replace the exotic plantations. Where the damage is more extreme, the resulting community may be a combination of native species and plantation trees (Lugo, 1992). Similar procedures were effective in Sri Lanka (Ashton et al., 1997) and Uganda (Fimbel & Fimbel, 1996) where Caribbean pine (*Pinus caribaea*) was used as a nurse plant to improve microenvironmental conditions enough that native tree species became established.

Despite the obvious and very important advantages of using non-native species to modify environmental conditions for subsequent establishment of native species, caution is advisable. To avoid creating major problems, the behavior of introduced species in their new environment should be well understood. Each tree species creates a unique environment that facilitates the development of a different native flora. In Costa Rica, a study of the role of tree plantations on the establishment and growth of native tree species found the quality and quantity of understory regeneration varied between plantation species, because each created a unique light environment (Guariguata *et al.*, 1995). Fast-growing fleshy-fruited trees create new habitat-forming islands in abandoned pastures because they attract fruigivores that bring new species (Nepstad, Uhl & Serro, 1991). Shape and foliage density influences a shrub's ability to ameliorate harsh wind and temperature regimes. Physical differences between shrubs influence their ability to facilitate the subsequent establishment of unplanted herbaceous species.

Differential species availability

The presence of a plant species following a disturbance depends on survival, successful migration, and establishment of its propagules (Gleason, 1939; Pickett *et al.*, 1987b). Many successional sequences probably reflect the limited number of species that were available after the disturbance, rather than a deterministic pattern of species replacement. The initial establishment of plants can have very long-lasting effects, particularly for trees (Roberts, 1987). Thus, the 'initial floristic composition' following a disturbance determines the outcome of succession, rather than each suite of species facilitating the entry of its successors (Egler, 1954). Thus, long-term vegetation changes are directed by the early introduction of certain species.

Directing differential species availability involves increasing the availability of desired species while reducing the abundance of less desirable species. While repair activities routinely include the artificial introduction of seeds or seedlings, they less frequently use subtle ecological strategies to manipulate species availability. Species availability may be addressed, in the long term, with repair strategies that attract vectors of desired seed, increase the capture of propagules carried in wind or water, and reduce the impact of seed predators. The proximity and spatial relationships of the various landscape components affect the effectiveness of these strategies. Thus, the overall landscape design becomes an important element.

DISPERSAL

Since seed dispersal rates influence the pace of successional change, sites located great distances from natural seed sources develop slowly (Ash, Gemmell & Bradshaw, 1994). Consequently, early successional species persist. Most repair programs artificially return desired species by planting seed and/or transplanting seedlings. However, artificial revegetation (artificially induced recovery) is expensive and commercial seed sources are available for relatively few species. While this may have highly significant benefits, it returns only a fraction of the natural diversity and functional processes. Consequently, under certain circumstances it is beneficial to enhance the attractiveness of the repair site to animals that disperse propagules of desired species. There is some potential for the use of domestic livestock to transport seed.

The quantity and quality of the seed dispersal process determines the effectiveness of seed transport mechanisms (Vander Wall, 1993). Although wind moves large amounts of seed, its dispersal is of relatively low quality. Wind-dispersed seed are influenced by prevailing wind directions, seed-trapping structures, and the relative topographic position of seed donor and seed receptor sites. Dispersal by seed-caching animals may move a high proportion of the seed into microhabitats that improve establishment success (Vander Wall, 1993). Scatter hoarders also provide a very effective means of seed burial.

Artificial perches accelerate the recovery of forest diversity in highly fragmented landscapes. Sites with artificial perches develop a greater abundance and diversity of bird-dispersed plants than sites without perches (McClanahan & Wolfe, 1993). Prior to implementing this strategy, you should determine which plant species the birds are likely to introduce. Birds, or other animals, may introduce invasive weeds that have the capacity to dominate a site and exclude other species (Robinson & Handel, 1983). A management scheme for controlling those species might be necessary.

Simply increasing seed immigration into disturbed sites may not overcome site deficiencies. McClanahan and Wolfe (1993) found that plant associations under perches did not reflect seed inputs and were not successful in reestablishing late-successional genera. Established plants were typically early-successional shrubs with small seeds. Late-successional tree species in the seed rain had greater mortality and were not present among the recruited genera. Severely disturbed sites did not provide suitable environments for late-successional species (McClanahan & Wolfe, 1993).

The relative abundance and ubiquitous distribution of livestock in most wildland ecosystems makes them important seed dispersers. There are numerous examples where livestock dispersed seed onto degraded sites after eating the desired seed (Burton & Andrew, 1948; Wilson & Hennessy, 1977; Wicklow & Zak, 1983; Ahmed, 1986; Brown & Archer, 1987; Simao-Neto, Jones & Ratcliff, 1987; Jones, Noguchi & Bunch, 1991; Barrow & Havstad, 1992; Gardner, 1993; Ocumpaugh, Archer & Stuth, 1996). Using livestock to spread seed requires an understanding of seed survival, rate of passage through the digestive system (a function of seed size, hardness, specific gravity, and animal diet), germination rates in dung, and subsequent seedling establishment success (Archer & Pyke, 1991). For example, soft-seeded species lose viability more quickly than hard-seeded species. These issues have been addressed to varying degrees for several plant and livestock species (Yamada & Kawaguchi, 1972; Jones & SimeoNeto, 1987).

Goats and sheep were so effective at establishing *Acacia tortilis* and *Prosopis chilensis* over great distances they were recommended for seeding semiarid areas (Ahmed, 1986). *Prosopis* and *Acacia* seedlings commonly emerge from the dung of ungulates on at least four continents (Archer & Pyke, 1991). The rapid spread of these hard-seeded legumes is testimony to the effectiveness of livestock as dispersal agents (Brown & Archer, 1987). Another study in the Chihuahuan desert fed gelatin capsules containing seed of four-wing saltbush (*Atriplex canescens*), alkali sacaton (*Sporobolus airoides*), blue panicgrass (*Panicum antidotale*), and sideoats grama (*Bouteloua curtipendula*) to steers (Barrow & Havstad, 1992). About 95% of the recovered seed passed through steers within 72 hours and total recovery was 10%, 46%, 62%, and 0% for fourwing saltbush, alkali sacaton, blue panicgrass, and sideoats grama, respectively. Germination of seeds recovered 48 hours after ingestion was 15%, 50%, 41% for fourwing saltbush, alkali sacaton, and blue panicgrass, respectively.

SEED PREDATORS

The selective consumption of seed by granivores affects the species composition of plant communities (Everett, Meewig & Stevens, 1978; Whitford, 1978; Kelrick & MacMahon, 1985; Kelrick *et al.*, 1986). This selectivity may profoundly influence the results of certain types of repair practices, particularly broadcast seeding. Rodents and ants have decreased seed reserves by 30–80% in some ecosystems (Archer & Pyke, 1991). Forb seed in an old-field community were removed at rates of 3–45% per day (Mittlebach & Gross, 1984). While rodents are the most important granivores of arid and semiarid regions of North America and Israeli, ants are more important in South America, Australia, and South Africa (Kerley, 1991). Patterns of seed removal by small mammals in the South African Karoo differ from those in North American and Israeli deserts, but resemble those in Australian and South American deserts (Kerley, 1991).

Small seedeaters, such as ants, influence plant densities because of their efficient foraging capability at low seed densities. Manipulative studies in the Sonoran (Davidson, Inouye & Brown, 1984) and Chihuahuan deserts (Davidson & Samson, 1985) found that removing harvester ants, which specialize in small seeds, increased the densities of annual plants. Vertebrate granivores exploit a range of seed types and sizes, but seem to prefer large seeds (Davidson, 1993; Heske, Brown & Guo, 1993). This indicates that seed predation will retard succession if large seeds are preferred or have little impact if small seeds are preferred (Davidson, 1993). The greater selectivity of large, vertebrate seed-predators can alter the composition of plant species and the direction of succession (Davidson & Samson, 1985; Samson & Phillippi, 1992). Experimentally excluding rodents that specialize in larger seeds reduced the diversity and productivity of ephemeral plants (Samson & Phillippi, 1992). Seed predation of large-seeded species retarded succession in the California chaparral (Louda, 1982). However, a study in the sagebrush steppe suggested that the percentage of soluble carbohydrates rather than seed size was the best predictor of rodent seed preference (Kelrick & MacMahon, 1985).

Selective seed removal can influence wildland repair efforts, particularly when seed are broadcast over the soil surface (Nelson, Wilson & Goebel, 1970). Kangaroo rats (*Dipodomys* sp.) in the Chihuahuan desert have more effect on vegetation than livestock because of seed predation, seed caching, and soil disturbance (Heske *et al.*, 1993). A study in the semiarid interior of Washington State made grass seed available 'cafeteria-style' and measured consumption by about 25 bird species that inhabited the area during typical reseeding periods (Goebel & Berry, 1976). Birds preferred perennial grass seed to annual grass seed. In arid and semiarid regions where broadcast seeding is commonly practiced, birds can inhibit establishment of desirable species through their diet selection (Goebel & Berry, 1976).

Several strategies have reduced seed predation, but the results were variable: (1) design sites with low perimeter-to-area ratios; (2) disperse seed at unpredictable times or when granivore populations are low; (3) supply more seed than predators can remove (predator satiation); (4) manage the habitat to increase populations of birds and mammals that prey on seed-eating animals; (5) apply rodenticides to reduce granivore numbers; (6) dye seeds another color to confuse granivores.

Although few repair projects are able to design large, round sites, or seed during low points in predator population cycles, these strategies are occasionally useful. Seeding adjacent sites in a single year will increase the perimeter-to-area ratio and reduce the potential for predator population expansions (Archer & Pyke, 1991). An alternative approach would be to stagger seeding efforts on adjacent sites to allow granivore populations to return to preseeding levels before seeding adjacent sites (Archer & Pyke, 1991). Sequential seeding of adjacent sites over several years may increase the number of seed predators.

Predator satiation is too expensive for most wildland situations, but is practical for certain situations. Supplying a 2:1 ratio of sunflower seed to lodgepole pine (*Pinus contorta*) seed reduced rodent consumption of pine seed from 85% to between 28% and 58% (Sullivan & Sullivan, 1982). In a separate study, Douglas fir seedling establishment improved from 5% to 50% after sunflower seed were added (Sullivan, 1979). Although the use of predator satiation during wildland repair operation is not widely tested and no general recommendations are available, it suggests intriguing possibilities.

Habitat management, rodenticides, and colored dyes have reduced seed losses in some situations. Habitat manipulations that increase the numbers of raptors and mammalian predators can reduce rodent populations. Developing nesting and perching structures for raptors (MacMahon, 1987) and designing hiding and approach cover for mammalian predators (Archer & Pyke, 1991) can reduce seed losses. Rodenticides and colored dyes reduced seed losses to rodents and birds after broadcast seeding (Vallentine, 1989). Rodenticides reduced rodent numbers. Colored dyes disrupted search patterns used by the granivores to locate certain seed. Changing the color of the seed may reduce the granivores' ability to recognize the seed as food. However, the granivores will probably establish a new search pattern (for dyed seed) if those seed are abundant.

5

Selecting plant materials

Introduction

Having decided to add plants rather than rely on natural recovery, we must choose between strategies that modify the site for the desired species, or strategies that rely on plants tolerant of existing conditions. In either situation, it is essential to select plants that improve resource availability, rather than simply exploiting and depleting them. Plants that improve harsh microenvironmental conditions are also important. Although the emphasis throughout this book is on adapted plants that use autogenic processes to modify conditions over time, some site modification is often required to initiate and direct natural recovery processes.

Selecting the appropriate species and species mixtures for each part of the landscape is a critical decision that has several important considerations (Table 5.1). Many projects were early successes and long-term failures because of poorly adapted plant materials. Well-adapted plants not only survive and reproduce under existing conditions; they are adapted to rare or infrequent events. Appropriate plant materials repair damaged hydrologic and nutrient cycling processes while ameliorating harsh microenvironmental conditions. The ability of plants to modify habitats led to their being viewed as 'ecosystem engineers' (Jones, Lawton & Shachek, 1994; Jones, Lawton & Shachek, 1997). Properly selected plant associations encourage positive interactions among organisms within and adjacent to the repair site. The process-oriented approach suggested here emphasizes repairing processes, with the return of particular species or species groups being a secondary

Table 5.1. *Considerations for selecting appropriate species and species mixtures for each part of the landscape*
Relevant chapters with additional information for each item are listed.

Set clear and achievable objectives (Chapters 1 and 8):
• for repairing damaged ecosystem processes (Chapters 2 and 3);
• for directing vegetation change (Chapter 4);
• that include the time frame for achieving repair objectives (Chapters 1 and 4);
• for goods and services (forage, timber, fodder, fuel-wood, soil conservation, water management, and water quality) (Chapters 1, 3, 4, and 8); and
• that consider ecologic, practical, and economic restrictions of potential plant materials (seed cost, availability, quality, and at proper time) and requirements for additional equipment (Chapters 1, 4, 5, 6, and 7).

Identify suitability of plant materials to the existing and anticipated:
• climatic conditions (including unusual events) (Chapter 4);
• successional stage (Chapter 4);
• microclimatic conditions (Chapter 4);
• soil status (water, biotic, compaction, nutrient, pH, erosional forces) (Chapters 2, 3, and 5); and
• disturbance regime (herbivory or fire), insects and diseases (Chapter 4).

Formulate species mixtures for each site that:
• meet management objectives for goods and services (Chapters 1 and 4);
• rapidly stabilize site and repair damaged processes (Chapters 2, 3, and 5);
• provide appropriate genetic and functional diversity (Chapters 2, 3, and 5);
• are compatible with adjacent landscape components and other species (Chapters 2, 4, 5, 7, and 8);
• improve site conditions (microenvironment, hydrology, soil and nutrients), initiate, and continue autogenic repair processes (Chapters 2, 3, 4, 5, and 6); and
• specify the use of seed, seedlings, or plant parts (Chapters 5, 6, and 7).

concern. While this approach does not preclude recreating historical ecosystems, it considers that goal secondary to repairing ecosystem function.

For each species, we must decide whether to plant seed, transplant seedlings, move entire plants or even parts of plants. We should always assess the practical availability of potential species before selecting them. Adequate plant materials must be available, at the correct time, and at an acceptable cost. Does a particular species require the purchase or rental of equipment? Does maintenance of a species or group of species require additional nutrients, basic cations, or protection from grazing animals? Are they worth the additional expense or trouble? Answer each of these questions carefully before selecting plant materials.

Species and species mixtures

There is growing evidence that our basic paradigms for breeding and selecting plant materials are seriously flawed when applied to wildland. We are selecting, and even breeding, plant materials for the wrong traits. For example, in semiarid and arid ecosystems, plant materials are often selected for forage quality and productivity, which are sometimes determined in irrigated nurseries. This often selects the wrong species and genotypes for both water- and nutrient-limited environments. The most productive genotypes are poorly suited to water-limited environments. A study of 29 *Agropyron desertorum* varieties found the most productive genotypes were less efficient in their use of water to produce dry matter (Johnson *et al.*, 1990). Thus, water use efficiency (WUE) was inversely proportional to productivity, suggesting the benefits of less productive genotypes in water-limited wildland ecosystems.

The most productive species and genotypes are poorly suited to low-nutrient environments. Species adapted to low-nutrient environments grow slowly, use nutrients conservatively, and produce herbage that is less attractive to herbivores and decomposer organisms (Grime & Hunt, 1975; Poorter & Remkes, 1990; Poorter *et al.*, 1990; Aerts & Peijl, 1993). These traits characterize plants of infertile environments; they use nutrients very efficiently (Flanagan & Cleve, 1983; Coley *et al.*, 1985; Coley, 1988; Bryant, Reichardt & Clausen, 1992; Hobbie, 1992; Aerts & Peijl, 1993). In contrast, the plants most often selected for wildlands grow rapidly and have high quality forage. These attributes accelerate nutrient cycling and increase nutrient losses. In infertile environments without continuing nutrient subsidies, these traits create an inherently unsustainable situation (Burrows, 1991; Myers & Robbins, 1991). Where we cannot afford high energy subsidies, we should match plant requirements to nutrient availability. Attempting to maintain early successional or high N demanding species in wildland sites with low N supply leads to disappointment (Burrows, 1991).

Although plant breeding has played a very important role in wildland repair efforts, it is somewhat controversial to some native plant proponents. Three conditions justify these intensive plant breeding programs (Jones, 1997). First, the species are relatively important, but limited by a specific problem. Second, less manipulative methods are incapable of addressing the problem. Third, the tools of plant breeding can solve the problem by selecting for traits elucidated by ecological, physiological, or genetic concepts.

Species are available for almost any environment, but the options are more limited in extreme environments. Harsh environments limit growth, and production potentials, thus increasing the time required to improve those conditions. Common species thrive in a variety of situations (i.e., wide ecological amplitude). They are used for at least three good reasons: (1) reliable establishment without site-specific research; (2) wide ecological amplitudes making them less susceptible to changing conditions; and (3) well-known propagation and establishment techniques (Coppin & Stiles, 1995). These traits resulted in standardized mixtures for many situations. We should always consider standardized mixtures that show reliable results and, regardless of the procedures used, assess the practical availability of the materials and any additional equipment or management inputs required for their use.

Wildland repair programs seek to provide necessary goods and services (soil conservation, water quality, water quantity, biological diversity, esthetics, forage, wood, wildlife habitat, etc.). How do we identify suites of plant materials to achieve those elusive objectives? Contemporary strategies for selecting wildland species mixtures emphasize (1) native species; (2) species diversity; (3) functional diversity; (4) assembly rules; and (5) the self-design capacity of ecosystems.

Native species

A prudent approach is to begin by considering native species for most wildlands. They have proven long-term climatic adaptations and their coexistence with other native species suggests compatibility. Native species are less likely to cause new problems, unless other problems (e.g., poor grazing management) exist. There are two critical issues relating to the use of native species. The first involves exactly what constitutes an appropriate 'native genotype.' The second and most contentious issue concerns circumstances for the use of nonnative species.

There are different opinions on plants or seed collected long distances from repair sites. One view is that seed from distant sources are dangerous because they are poorly adapted and contaminate the local genetic material, reducing the vigor and competitive ability (Knapp & Rice, 1994). It is suggested that native seed should come from less than 330 km north or south and 160 km east or west of the repair site (Welch, Rector & Alderson, 1993). Others argue that collecting seed more than 100 m (herbaceous plants) to 1 km (woody plants) from the planting site is dangerous because it risks 'genetic pollution' (Linhart, 1993).

Genetic pollution refers to introgression, which involves hybridization followed by backcrossing and the fixing of those backcrosses. Genetic pollution dilutes the gene flow of native populations and creates hybrids that are poorly adapted to local conditions (Linhart, 1993). However, we might also view these 'problems' as natural selection, where the unfit do not persist. These hybrids are occasionally more fit and better adapted to local conditions. Strict distance requirements are neither practical nor supported by genetic or evolutionary evidence. Introgression is widespread in wildland plants and is an important evolutionary mechanism. Plants derived from one taxon, but with introgressed exotic genetic material, are more likely than F_1 hybrids to be fertile and adapted to some ecological niche. Introgression creates genetic variation that occasionally generates new genetic material that is better adapted.

Species with large geographic ranges typically have more genetic variability than narrow endemics. Long-lived, woody species are more variable than short-lived species (Linhart, 1993). Strongly outcrossed, wind-pollinated species have fewer differences between populations than self-pollinated species (Linhart, 1993). Even relatively local conditions have a strong influence on the genetic variability of some species. Pronounced differences in exposure, soil conditions, or even community structure can produce unique genetic compositions. Strong, local genetic differences do not develop in (1) most aquatic species; (2) species that underwent serious population bottlenecks; and (3) species with high phenotypic plasticity (Linhart, 1993).

Multiple-source introductions use a mixture of genotypes collected from a larger portion of the species' range. This strategy was combined with plant breeding approaches to develop the convergent–divergent approach for native plant materials (Munda & Smith, 1993). The convergent–divergent breeding strategy selects for widely adapted cross-pollinated materials by practicing artificial recurrent selection on a multiple-origin polycross. First, multiple-origin materials, from a defined ecological area (convergent phase), are intermated (polycross). Second, intermated populations are established at various sites within the defined ecological area. This provides selection opportunities at many sites within the targeted ecological area (divergent phase). Third, plants are returned from all locations to a common site for a second intermating (second convergent phase). Forth, this process is repeated to produce widely adapted genetic materials (at least within the defined area) (Jones, 1997).

When locating plants for harsh environments, it is best to collect seed from similar sites. Many seedings in arid and semiarid ecosystems fail because the seeds (or transplanted seedlings) are poorly adapted. Commercial seed sources seldom provide enough seed source information. Thus, informed seed collectors that seek conditions similar to the repair site are preferred (Van Epps & McKell, 1978). They can select from among natural genotypes to find those that appear most capable of growing on the repair site. The collection and use of site-specific plant material is appropriate for four conditions: (1) specific ecotypes are needed; (2) commercial sources are inadequate; (3) there is high potential for immediate use; and (4) commercial potential is limited (Jones, 1997).

USING NONNATIVE SPECIES

Do nonnative species have a role in the repair of damaged wildlands? If so, under what conditions are nonnative species useful? Native species should be returned to sites where they are well-suited and meet management objectives. However, there are circumstances where native species are poor choices: for example, where previous disturbances have substantially altered site conditions so that native species are no longer able to establish and persist. Social or economic circumstances may preclude the use of native species that do not provide the necessary goods or services. In these, and other situations, we need to consider nonnative species or a mixture of native and nonnative species. It is possible that nonnative species have a role to play in restoring functional aspects of ecosystems, but this must be related to the overall goals of restoration and the context within which restoration is carried out (Hobbs & Mooney, 1993).

We must differentiate between problematic nonnative species and those that might fill essential functional or economic roles. Ecosystems are routinely bombarded by new arrivals, without altering large-scale ecosystem properties and processes in any meaningful way (Vitousek, 1990). However, most ecosystems also have nonnative species that have caused extreme disruptions (Mack, 1981; Vitousek, 1990; Lodge, 1993; OTA, 1993; Lonsdale, 1994; Cronk & Fuller, 1995), often by dramatically altering ecosystem processes (D'Antonio & Vitousek, 1992). Some of the worst weeds in the United States were introduced by people who thought they would be beneficial: kudzu (*Pueraria lobata*), water hyacinth (*Eichhornia crassipes*), and multiflora rose (*Rosa multiflora*) (OTA, 1993). Kudzu was widely promoted for erosion control in the 1940s; yet the same characteristics considered beneficial (rapid growth, ease of propagation, and wide adaptability) allowed it to become a serious problem throughout the southeastern United States (OTA, 1993). Purple loosestrife (*Lythrum salicaria*) is an attractive nursery plant, but a major wetland weed. The melaleuca tree (*Melaleuca quinquenervia*) is rapidly degrading Florida Everglades wetlands by outcompeting indigenous plants and altering topography and soils (OTA, 1993). In Australia, the nonnative rubber vine (*Cryptostegia grandiflora*) forms dense thickets that smother vegetation 30–40 m high. Since it is very difficult to predict where new species will cause problems (Mack, 1996), we should be extremely careful when introducing nonnative plant materials. Thus, our use of nonnative species should be limited to regions with established populations of the nonnative species that have not created problems.

Some nonnative species are well adapted to certain sites and are even superior for some management objectives. Crested wheatgrass (*Agropyron cristatum* and *Agropyron desertorum*) and some of the African lovegrasses (*Eragrostis lehmanniana*, and *Eragrostis curvula* var. *conferta*) are widely used in the western US because of their ease of establishment and persistence under arid and semiarid conditions.

They tolerate grazing, establish on disturbed sites, persist where planted, and spread into surrounding native vegetation (Hull & Klomp, 1966; Hull & Klomp, 1967; Marlette & Anderson, 1986). Crested wheatgrass inhibits the development of more diverse plant communities (Marlette & Anderson, 1986). In the semidesert grasslands of southeastern Arizona, areas seeded to *Eragrostis* species support fewer indigenous plant and animal species than areas dominated by native perennial grasses, even 20 years after livestock removal (Bock *et al.*, 1986). *Eragrostis lehmanniana* now dominates at least 145 000 ha in southern Arizona (Anable, McClaran & Ruyle, 1992). Compared with fully functional native grasslands, *Eragrostis*-dominated areas are species poor. The use of African lovegrasses and crested wheatgrass will continue, because they establish more readily than other species. The same characteristics that make them desirable for some applications cause problems where species richness is an important objective.

The apparent compatibility of coexisting native species does not prove they coevolved with each other – a commonly stated reason for their use. Some scientists believe no evidence exists 'that successful, productive communities consist mostly of species that evolved together and developed complex mechanisms for co-existing in a delicate balance.' They argue that no special significance should be placed on native species and that 'the sanctity attributed to climax vegetation because it is natural, repeatable, and stable in species composition is without merit' (Johnson & Mayeux, 1992).

Some nonnative grasses have long-term negative impacts that degrade both the soil and vegetation. In North America, the conversion of native, mixed prairie to monocultures of either crested wheatgrass or Russian wildrye (*Elymus junceus*), decreased root mass, soil organic matter and the monosaccharide content of dry soil aggregates (Dormaar *et al.*, 1995). In arid and semiarid environments, crested wheatgrass stands may have 10 times more bare ground than native communities (Lesica & DeLuca, 1996). Despite covering less ground, crested wheatgrass produces more aboveground biomass and less belowground biomass than native species. Less belowground growth produces less root detritus and root exudates to feed microbial activities that stabilize soil aggregates (Lesica & DeLuca, 1996). Crested wheatgrass stands have higher bulk density, fewer water-stable aggregates, and less organic matter or nitrogen compared with native grasslands (Biondini, Klein & Redente, 1988). Crested wheatgrass supplies soils with more carbohydrates and less organic nitrogen (Klein *et al.*, 1988) – about half the organic nitrogen supplied by native grasses. This increases soil organic nitrogen mineralization (i.e., the so-called priming effect), which increases the net soil organic nitrogen demand (Lesica & DeLuca, 1996). Productivity increases until the organic nitrogen pool is consumed. Under these circumstances, certain introduced grasses cannot maintain the soil's chemical or physical quality.

Species diversity

Species selection strategies that emphasize diversity assume species-rich ecosystems are more stable and less susceptible to damage from unusual climatic events, disease, or insects. It is widely believed that mature communities have numerous niche-differentiated species that complement each other rather than directly competing with one another (Whittaker, 1970). There is also compelling evidence that biodiversity confers ecosystem stability by buffering against natural and artificial perturbations (Smith, 1996; Tilman, 1996). However, species diversity alone does not necessarily provide that protection.

Much of the stability, resistance to damage, and self-repairing capacity associated with species-rich ecosystems probably results from functional diversity rather than species diversity. Although there is some evidence linking functional complexity with short-term ecosystem stability (Van Voris *et al.*, 1980), functional complexity is not necessarily obtained by adding species. A function can be performed by several species and a particular species might perform multiple functions. Thus, there is no one-to-one relationship between species and ecological functions (Haila, Saunders & Hobbs, 1993). While it is unclear whether proper ecosystem functioning requires biodiversity, there is no doubt that proper functioning is necessary to maintain biodiversity (Hobbs, 1992b). The challenge for ecological repair is to return and maintain ecosystem processes in fragmented and extensively modified systems. Adding species that perform essential functions will contribute to functional integrity. Genetic diversity plays an important role in determining both immediate success and adaptability to future change.

GENETIC DIVERSITY

The selection and development of improved plant varieties has probably received more research attention than any other aspect of wildland repair. Although selecting the correct variety improves success, reduced genetic variability can limit success under changing or unanticipated conditions. This is an important distinction because of the inherent variability and unpredictability of most wildland ecosystems. The high genetic variability in many wildland species is a necessary adaptation to unpredictable climates, and variable soils and topographic conditions (Stutz, 1982). Although poorly understood, well-adapted varieties are most effective when placed into relatively uniform conditions matching their original home. In contrast, when planting into highly variable sites, it may be desirable to use plant materials with high genetic diversity (Stutz & Carlson, 1985) to increase the likelihood that at least some individuals will establish and succeed on the site. However, it has also been suggested that maintaining maximum genetic diversity inhibits the evolution of adaptations to local conditions (Guerrant, 1996).

In the absence of materials known to be well adapted to existing conditions, the preponderance of evidence still supports the value of high intraspecific genetic diversity. We increase genetic diversity by (1) using seed from populations with high natural diversity or (2) mixing seed from multiple sources. Each of these strategies for selecting intraspecific genetic diversity has both proponents and critics. Proponents of single-source gene pools believe that maintaining the integrity of lineages specifically adapted to local conditions is of primary importance because crossing with 'foreign' genetic materials leads to outbreeding depression (Guerrant, 1996). They also use philosophical arguments against 'contaminating historically mixed lineages'. It is now believed that intrapopulation variability in the total amount of DNA effectively enables individuals to exploit different temporal and microclimatic niches within a diverse habitat (Mowforth & Grime, 1989).

Multiple-source materials combine a species' genetic diversity within an individual site or ecologically similar sites (Jones, 1997). This approach provides an opportunity to reduce risks. Bulk mixtures of self-pollinating accessions amalgamate material from a variety of sites collected from a common area. At least part of the material is adapted to the seeding site. Proponents believe multiple-source introductions provide the genetic diversity that allows species to adapt to future conditions. Commercial seed companies now offer composite collections of up to 15 genotypes (different accessions or similar species) raised and sold as a mixture. However, if pooled seed production for these diverse genotypes occur under irrigation and fertilization, the diversity of the resulting seed crop is suspect. Almost any approach to pooled seed production selects for a subset of the entire group, rather than the whole group. This concern led to the development of seed production strategies for both cross-pollinated and self-pollinated species that seek to maintain or even increase genetic diversity of plant materials (Munda & Smith, 1993).

Pragmatic arguments for multiple-source introductions recognize the difficulty in locating sufficient seed for a specific site. For many species, only small quantities of seed are available from any single source and adaptations are unknown. This is problematic when we consider that natural genetic variability has been demonstrated on a scale of tens of meters or less for herbaceous species and 100–300 m for woody species (Linhart, 1993; Knapp & Rice, 1994). The practicality of using specific gene pools decreases as the size of the repair site increases.

An important source of genetic variation arises from chromosome races, resulting from either chromosomal rearrangements or polyploid increases in chromosome number (Stutz, 1982; McArthur, 1991). Chromosome races usually correspond to different habitats and exist for many wildland species. Although seldom understood, that information is becoming available for some species (Stutz, 1982; McArthur, 1991). For example, mountain big sagebrush (*Artemisia tridentata* ssp. *vaseyana*) tetraploids apparently derive *de novo* from diploids, but are better adapted to drier habitats (McArthur *et al.*, 1998).

The type of plant breeding system has a major impact on gene flow rates and the capacity for local adaptation (Rice & Knapp, 1997). Gene flow is the transfer of genetic material from one population to another, such as from seeded nonlocal genotypes to the remnant native plants. Wind-pollinated plants have more rapid gene flows than do animal-pollinated species (Jones, 1997). Self-pollinated plant species have very little gene flow. Gene flow rates in highly outcrossing species can be strong enough to override the influence of natural selection. Thus, strong gene flow can reduce the ability of natural selection to create locally adapted populations, even across strong environmental gradients (Rice & Knapp, 1997). Under other conditions, gene flow may generate desirable new genetic combinations (Jones, 1997).

It is possible that landscape stability is related to the maintenance of some minimum level of genetic diversity (Linhart, 1993; Knapp & Rice, 1994; Urbanska, 1995). Landscapes that do not allow for the continuing evolution of species and populations may restrict the adaptive ability of future generations (Frankel, 1974). Species are only able to adapt to changing conditions if they have sufficient genetic variability to allow natural evolutionary processes to select adaptive traits (Harris, 1984).

Functional diversity

Are complex species mixtures necessary for fully functional ecosystems? If energy and materials fluxes are the criteria for functionality, the mix of species may not be especially important (Ewel, 1997; Hooper & Vitousek, 1998). Many species-poor ecosystems maintain biotic controls over limiting resources. However, when trophic interactions, symbioses, pollination, and nutrient cycling are considered, then the exact mixture of species becomes important, with diverse communities containing more complex processes than simple communities (Ewel, 1997). It is increasingly, though not universally, believed that ecosystem function can be restored without using the same species (Ewel, 1986; Westman, 1990; Ewel, Mazzarino & Berish, 1991; West, 1993). However, it is important to remember that we usually assign plants (or other organisms) to functional groups based on individual attributes or their responses to specific kinds of events. This application of 'single-factor ecology' is too simplistic to accurately describe how species function within a particular ecosystem.

No functions are unimportant, but some address known limitations of the site and/or climate and deserve emphasis. For example, functional groups that improve hydrologic processes (e.g., infiltration, erosion, or soil structural development), water use efficiency, or microenvironmental conditions are critical in water-limited environments. Sites with serious nutrient depletion benefit from species that increase the uptake, capture, retention, and use-efficiency of the most depleted nutrients. This emphasis on repairing damaged primary processes (hydrology, nutrient cycling, and energy capture) allows other functions to either develop naturally or be returned later through artificial means.

Since we cannot include all functional groups (functional guilds) into each seeding mix, we focus on functional groups that regulate limiting or dominant ecosystem processes (Walker, 1992). Species affect ecosystem processes with traits that (1) modify the availability, capture, and use of soil resources; (2) affect feeding relationships (trophic structure) within a community, and (3) influence the frequency, severity, and extent of disturbances such as fire (Vitousek, 1990; Chapin *et al.*, 1997). We assess the important functional grouping by considering (1) ecological strategies; (2) regeneration strategies; (3) pollination requirements; (4) contributions to site stabilization and primary process repair; and (5) functional redundancy.

ECOLOGICAL STRATEGIES

An ecological strategy is 'a grouping of similar or analogous genetic characteristics which recurs widely among species or populations and causes them to show similarities in ecology' (Grime, 1986). Although the term strategy has teleological implications that some find disturbing (Burrows, 1991), ecological strategies provide a compact framework for synthesizing ecological information.

The success of a species depends on its strategies for establishment and growth. Plant growth and reproduction are adversely affected by factors that reduce growth (stress) or remove biomass (disturbance) (Grime, 1977; Grime, 1979). Stress results from external factors (e.g., low light, low water, low nutrients, or temperature extremes) that limit plant growth, while disturbances (e.g., fire, herbivory, or physical damage) remove plant biomass. Obviously, stress and disturbance occur in various combinations of intensity and/or frequency, but only four combinations occur if we only consider low and high levels. Of these four potential combinations, only three are viable, since the combination of high stress and high disturbance results in mortality. Plants within each of the remaining three groups (stress-tolerant, competitive, and ruderal) have similar adaptations and predictable patterns during succession (Table 5.2). These categories describe plant response to a particular set of environmental conditions, such as resource availability and disturbance regime, rather than inherent properties of the plants themselves (Smith & Houston, 1989). Generalizations produced by strategy theory help us select species with characteristics that enhance their survival. There is no viable strategy for high stress and high disturbance, conditions all too common in severely degraded wildlands. In these situations, we must remove or reduce the effects of either stress or disturbance (Coppin & Stiles, 1995).

Severe stress is an inevitable feature of severely damaged environments. Stress is also induced by other plants (e.g., dense shade or the sequestration of mineral nutrients) (Grime, 1986). Stress tolerance differs from competitive stress in that one or more stresses operate almost continuously throughout the year and affect all species. These environments provide little opportunity for stress avoidance through morphological adaptations or seasonal growth patterns (Grime, 1986). Consequently, stress tolerators have characteristics that insure the survival of mature individuals under harsh environments by reducing allocations to vegetative growth and reproduction (Table 5.2). They are well adapted to unproductive environments with strong environmental stresses (Grime, 1977; Grime, 1979).

Competitors dominate competitive environments with abundant resources and little disturbance (Table 5.2). They devote significant resources to vegetative growth, in contrast to stress-tolerant plants that grow slowly and develop long-lived structures (Grime, 1977; Grime, 1979). These conditions lead to a dense cover of perennial plants that captures resources rapidly to construct new leaves and roots (Table 5.2). Competitors are most effectively used in mature ecosystems or damaged ecosystems with the potential for rapid vegetative development.

Disturbance in more fertile environments promotes ruderals with rapid growth, short life spans, and prolific reproduction (Grime, 1986). Ruderals are early successional herbaceous species with short life spans and high seed production (Table 5.2) and are most common on disturbed environments with high productive potential (Grime, 1977; Grime, 1979). In disturbed environments, properly selected ruderal species hasten site modifications necessary for the eventual dominance of more desirable species. Since disturbed sites may be ecologically young, climax species may not establish, survive, or persist on those sites. If that difficulty is anticipated, early seral species (e.g. vigorous annual, biennial, and/or short-lived perennial species) can be included on sites with poorly developed soils, even though they may not persist (DePuit, 1988b). Properly used, ruderal or pioneer species may accelerate successional processes by (1) stabilizing the soil; (2) increasing soil organic matter; (3) enriching soil nutrients; and (4) competitive exclusion of less desirable pioneer species.

Table 5.2. *Comparison of (a) the attributes of competitive, stress-tolerant, and ruderal plants and (b) possible uses for those traits*

	Competitive	Stress-tolerant	Ruderal
(a) Attributes			
Adapted to conditions of	Intense competition	Abiotic stress	Disturbance
Life forms	Herbs, shrubs, trees	Lichens, herbs, shrubs, trees	Herbs
Longevity	Long or relatively short	Long to very long	Very short
Flowering frequency	Usually each year for established plants	Intermittent flowering over long lifetime	High flowering frequency
Proportion of annual production allocated to seed production	Small	Small	Large
Maximum relative growth rate	Rapid	Slow	Rapid
Photosynthesis and uptake of mineral nutrients	Seasonal, during long continuous period of vegetative growth	Opportunistic, may be uncoupled from vegetative growth	Opportunistic, coinciding with vegetative growth
Leaf phenology	Well-defined periods of maximum potential production	Evergreens, with variable leaf production patterns	Short phase of leaf production with high production potential
Leaf type	Robust, often mesomorphic	Often small, leathery, needle-like, or succulent	Variable, usually mesomorphic
Storage of photosynthate and nutrients	Rapidly incorporated into vegetative structure but a portion is stored to fuel next season's growth initiation	Storage in leaves, stems and/or roots	Storage in seeds

	Medium	High	Low
Ratio of root to shoot mass	Medium	High	Low
Longevity of leaf and root tissue	Relatively short	Long	Short
Luxury nutrient consumption	No	Yes	Yes
Mortality	Density dependent	Density independent	Density independent
Response to stress	Rapid allocation to vegetative growth	Slow response of small magnitude	Rapid shift from vegetative growth to flowering and seed production

(b) Potential roles in repairing damaged wildlands

Medium	High	Low
Sites with mature vegetation and intense competitive pressure (e.g., densely vegetated grasslands)	Sites with harsh, unrelenting abiotic environment	Sites that have been highly disturbed sites or have short intervals between disturbance events
Interseeding into established vegetation	Deserts and other ecosystems with severe water limitations	For rapid, short-term site stabilization following removal of vegetation
Densely vegetated wetlands, forests, grasslands, or shrublands	Nutrient-poor sites	Sites with short, unpredictable growing conditions
Sites with the potential for rapid vegetative development	Very acid soils	
	Densely shaded forest floors	
	High elevation sites with harsh UV light exposures and temperature extremes	
	Soils with high levels of sodium or other toxic materials	

Source: Grime (1979, 1986).

Most ecosystems have pulsed inputs of water and nutrient availability. Plant growth and resource availability are synchronized in fully functional ecosystems. We match growth and resource availability by selecting plants with (1) seasonal growth patterns that match resource availability; (2) different growth forms (above and below ground); and (3) different responses to expected disturbances. Different seasonal growth patterns insure that plants acquire the periodically available nutrients and water. Therefore, while cool-season plants reduce leaching losses during wet winters, it makes no sense to plant them in an environment with only summer precipitation.

REGENERATION STRATEGIES

Plants evolved numerous regenerative strategies as adaptations to particular environments (Lovett Doust, 1981; Lovell & Lovell, 1985; Grime, 1989; Kotanen, 1997). Although most plants reproduce by seed, vegetative strategies for exploiting new areas provide another means of placing plant materials into meaningful ecological groups (Table 5.3). Regenerative strategies should be carefully considered before trying to direct successional processes (Grime, 1986). Plant strategies for regenerating and invading new areas are useful in developing species mixtures. Regenerative strategies are important because they determine the extent to which the vegetation can repair itself following damage (Coppin & Stiles, 1995). They help us determine which species have the potential to dominate sites (Table 5.3).

POLLINATION REQUIREMENTS

Pollination requirements of wildland plants are seldom considered. That is usually appropriate, since our emphasis is on primary processes and the most commonly used species are wind pollinated. Wind pollination is common in gymnosperms, Poaceae, Cyperaceae, Juncaceae, Chenopodeaceae, Polygonaceae and sporadically in many other families. However, pollination requirements are important in certain situations. Animal-pollination commonly involves insects, but birds and mammals are important pollinators for certain plants. Pollinator visitation is essential for the long-term survival of many species. Although many pollinators are relatively common and/or highly mobile, repair efforts may suffer from a lack of pollinators. This is most likely when

Table 5.3. *Common terrestrial regeneration strategies, functional features, and potential applications for those traits*

Strategy	Functional features	Conditions where strategy is advantageous
Vegetative expansion	New shoots have vegetative origin and remain attached to parent until well established	Productive or unproductive habitats with low disturbance levels
Phalanx	Tightly packed advancing front of ramets or tillers that restrict other plants from entering their clonal territory	Stable habitats with long disturbance-free periods
Guerilla	Prostrate plants with relatively long internodes on rhizomes or stolons that allow plants to rapidly exploit adjacent open areas	Where disturbances (i.e., fire, herbivory, soil disturbance) are of light to moderate intensity or spatially heterogeneous
Seasonal regeneration in vegetation gaps	Independent offspring (from seed or vegetative origin) produced in a single cohort	Seasonally predictable climatic or biotic disturbance regimes
Persistent seed or spore bank	Dormant seeds or spores present throughout year with some persisting more than 1 year	Spatially predictable, but temporally unpredictable disturbance regimes
Numerous widely dispersed seed or spore	Numerous offspring produced from widely dispersed (primarily by wind) seed or spore that are of limited persistence.	Sites that are relatively inaccessible or with spatially unpredictable disturbance regimes.
		Large sites and/or situations where seed are to be used in small scattered patches in hopes they will spread naturally over time
Persistent juveniles	Offspring produced from independent seed or spore and capable of long term persistence in juvenile stage	Unproductive sites with low disturbance intensities

Source: Adapted from Lovett Doust (1981), Lovell & Lovell (1985), Grime (1989), Kotanen (1997).

the repair site is small and populated by species not found in the adjacent vegetation (Majer, 1989). For example, the germination percentage of royal catchfly (*Silene regia*), a perennial prairie plant, from populations of greater than 150 individuals was high. Small populations had greater variation within and between populations. This reduced seed viability is caused by (1) inbreeding depression in recently reduced populations; or (2) increased proportions of inferior seed caused by reduced hummingbird visitation (Menges, 1991).

SITE STABILIZATION AND PRIMARY PROCESS REPAIR

Properly selected plants stabilize sites and repair damaged primary processes by (1) adding aboveground obstructions (that capture soil, nutrients, organic materials and propagules moving in wind or water); (2) increasing resource retention; and (3) being compatible with the inherent nutrient cycling regime. Numerous strategies are available that may be used to reduce nutrient loss, acquire more nutrients, increase soil pH, or use other sources of nutrients (Table 5.4).

Plant physical properties determine their effect on site stability (Table 5.5). Slope stability increases with plants that (1) develop a uniform cover (at least 70%) close to the soil surface; (2) have a dense laterally spreading root system; (3) grow rapidly; (4) have maximum impact during highest rainfall intensities; (5) resist mechanical damage; and (6) produce abundant litter that decomposes slowly (Morgan & Rickson, 1995c). Root structure is important for slope stability and aboveground vegetation has more influence on surface erosion processes. Strong taproots stabilize slopes against mass failures, while dense, lateral root systems increase the strength of the surface soil through cohesion (Styczen & Morgan, 1995). Potentially useful species should be evaluated for these vegetative traits. Slopes with erosion problems may require stress-tolerant species (Coppin & Stiles, 1995).

Wind erosion can be reduced with plants that lower wind velocity near the soil surface. Plants influence wind erosion in five ways: (1) foliage reduces wind velocity; (2) foliage traps moving sediment; (3) vegetation cover protects the soil; (4) root systems increase the resistance of the soil to displacement; and (5) vegetation controls soil moisture through shading, uptake, and transpiration (Morgan & Rickson, 1995c).

Table 5.4. *Strategies for selecting vegetation to improve nutrient and pH conditions*

Site conditions	Strategy alternatives	Features of function or strategy and sources of relevant information
Environments with low soil nitrogen or where additional N might facilitate development of desired species	Nitrogen-fixing associations	Ability to use atmospheric nitrogen (e.g., legume–*Rhizobium* associations, bluegreen algae, or associations with N-fixing *Actinomycetes*) (Lawrence *et al.*, 1967; Garcia-Moya & McKell, 1970; Tiedemann & Klemedson, 1977; Jeffries *et al.*, 1981; Langkamp & Dalling, 1983; Marrs *et al.*, 1983; Vitousek *et al.*, 1987; Myers & Robbins, 1991; Aronson, Ovalle & Avendaño, 1992; Reiners *et al.*, 1994)
Acidic soils	Cation-pumping species	Harvest basic cations from large volume of soil and concentrate them in surface soil (Zinke & Crocler, 1962; Alban, 1982; Kilsgaard *et al.*, 1987; Choi & Wali, 1995)
Phosphate- or water-limited environments	Mycorrhizal associations	Roots infected with mycorrhizae increase P uptake in P limited environments (Reeves *et al.*, 1979; Janos, 1980; Trappe, 1981; Chiarello *et al.*, 1982; Fleming, 1983; Read *et al.*, 1985; Borchers & Perry, 1987; Miller, 1987; Newman, 1988; Allen, 1989; St. John, 1990; Harper, Jones & Hamilton, 1992; Anderson & Roberts, 1993; Herrera *et al.*, 1993; Haselwandter & Bowen, 1996)
	Fine-scale highly branched root architecture without extensive elongation	Acquire resources from a larger volume of soil (Savill, 1976; Harper *et al.*, 1992; Toky & Bisht, 1992)
Environments with severe nutrient limitations	Slow-growing evergreen woody species	Low nutrient absorption capacity to reduce nutrient demand. Long tissue life and slow decomposition rate slow nutrient cycling and conserve it in the system (Bryant *et al.*, 1992)
Productive environments favorable for rapid growth	Rapid growing herbaceous or woody species	Rapid resource capture to support fast growth in competitive environment (Grime, 1977; Chapin, 1980)

Table 5.5. *Functional requirements and desired qualities of vegetation used to stabilize slope and control erosion.*

Function	Desired qualities of vegetation	Principal considerations
Soil reinforcement and enhancing soil strength	Maximum root development to the required depth (i.e. below the slip surface)	Deep-rooting species Anchorage Suitable soil profile conditions
Removing water from soil	Vigorous root development throughout soil volume Large transpiration area (i.e. leaf surface area)	Vigorous rooting species Substantial top growth throughout year Soil water balance Salt content and tolerance
Protecting soil surface from traffic	Vigorous development at soil surface of both roots and shoots Capacity for rapid self-repair	Species selection, short growth habit Management Soil fertility Inherent ability of soil to withstand traffic Use of reinforcing materials
Protecting surface from wind and water erosion	Vigorous root and shoot development at soil surface Rapid establishment Uniform cover density	Erosion risk Behavior of vegetation under high flow conditions Soil surface conditions Species selection Use of reinforcing materials

Reinforcing banks and channels	Ability to grow in wet conditions perhaps with variable water levels	Species selection with respect to ecological preference
	Rapid effectiveness	Growth habit
	Root reinforcement	Management
	Top growth absorbs wave impact	Use of reinforcing materials
	Low hydraulic roughness under high flow	
	Self-repair capacity	
Providing shelter or screening	Top growth of suitable height and/or density	Species selection
	Rapid development	Density of foliage
		Structural arrangement

Source: Adapted and modified from Morgan & Rickson (1995a).

FUNCTIONAL REDUNDANCY

Functional redundancy is the extent to which species within a functional guild are interchangeable. Some degree of redundancy is critical for the maintenance of essential functions. Functional redundancy reduces the chance of total system failure by providing backups in case of species loss (Hobbs, 1992a). Although functional redundancy appears to reduce the importance of species composition, redundancies within a guild should not be used as an excuse to reduce any species (DeLeo & Levin, 1997). Two species may perform similar functions, but inevitably differ in other respects. Thus, since we will never fully understand all the functions of species, prudence suggests we incorporate more 'functional redundancy' than would be necessary if we fully understood the functional roles of species. Functional redundancy helps decide which kinds of species will contribute the most to proper ecosystem functioning (Hobbs, 1992a; Walker, 1992).

The ability of a single genotype to express different phenotypes is phenotypic plasticity. Where conditions change more rapidly than the generation time of a species, phenotypic plasticity can be adaptive (Rice & Knapp, 1997). Phenotypic plasticity can increase functional redundancy when one species alters its function as it adjusts to the loss of another species. The greater this capacity to compensate for the loss of another species, the more functionally redundant are the species (Walker, 1992; Frost *et al.*, 1995; Johnson *et al.*, 1996). Plasticity is believed to be inversely correlated with genetic heterozygosity (Knowles & Grant, 1981), particularly where the environment is highly variable (Geber & Dawson, 1993). Therefore, it is quite possible that high phenotypic plasticity within a species will increase functional redundancy.

Assembly rules

The loss of some species is functionally undetectable, but the loss of others has serious impacts. Thus, species are viewed as either ecological 'passengers' or as ecological 'drivers' (Walker, 1992). Passengers have little community-wide impact and drivers have strong effects on communities. This is an oversimplification, but it reflects our understanding that some species (or perhaps functional guilds) are more important than others are. If so, we should initially focus on returning the 'drivers' to damaged ecosystems. With a more complete understanding of these relationships, we could develop ecosystem assembly rules to guide repair efforts. Assembly rules describe our view of how species assemble into communities (Diamond, 1975a). Evidence is accumulating that suggests assembly rules are stronger where resources are most limited (Wilson, Peet & Sykes, 1995b), which is usually the case in dysfunctional wild-lands.

Ideally, assembly rules specify which traits and functions (i.e., functional guilds) should occur in a particular environment and the best sequence of their introduction. With that knowledge, we could reassemble damaged ecosystems with a series of building blocks (species or their functional equivalents). We should begin by deciding where and under what circumstances important guilds should occur. Since we know ecosystem development is not completely deterministic, it is unlikely assembly rules will become very precise in the future. Similar communities contain similar species groups, so there are probably rules (of unknown precision) that direct that development. This is most evident in wetland communities, since they are largely deterministic and predictable, but still have strong stochastic elements in their development (Weiher & Keddy, 1995). While exact assembly rules may not be possible, assembly rules have been described for wetlands (Keddy, 1992; Weiher & Keddy, 1995), New Zealand rain forests (Wilson, Allen & Lee, 1995a), desert rodent communities (Fox & Brown, 1993), and a lawn community (Wilson & Roxburgh, 1994). However, the current generation of assembly rules, at least for terrestrial ecosystems, is not very useful for ecosystem repair efforts.

Continuing development of assembly rules will almost certainly require a better understanding of functional guilds, ecosystem drivers, keystone species, and critical link species. Species without functional equivalents are considered keystone species (Westman, 1990). Legumes or other species with nitrogen-fixing associations are widely viewed as keystone species. A shrub in the Namib Desert (*Acanthosicyos horridus*), is considered a keystone species because it binds the sand and improves the microenvironment (Klopatek & Stock, 1994). These changes facilitate the establishment and growth of other species. Plants are not the only keystone species. In the Chihuahuan–Sonoran desert ecotone, kangaroo rats (*Dipodomys* spp.) encourage shrub establishment to the degree that without kangaroo rats, the area changes from a shrub steppe to grassland (Brown &

Heske, 1990). The rodents' digging activities are believed necessary for shrub establishment, since grass outcompetes shrubs in the absence of kangaroo rats. Keystone species may be rare in natural communities or they may be common and seldom recognized (Krebs, 1985). This concept was extrapolated to an Extended Keystone Hypothesis which suggested that 'all terrestrial ecosystems are controlled and organized by a small set of key plant, animal, and abiotic processes that structure the landscape at different scales' (Holling, 1992).

Critical link species play vital roles in ecosystem function even though they are not easily noticed and form a small part of the total biomass (Westman, 1990). Critical link species may, or may not, be keystone species. The best example of critical link species are mycorrhizal fungi (West, 1993) that trade carbon fixed by vascular plants for improved phosphorus and water uptake. Their absence severely retards vascular plant recovery on some disturbed sites.

The goal of assembly rules is to restore essential ecological functions and encourage the biological interactions that contribute to the autogenic development of damaged ecosystems. At best, current assembly rules identify critical functional roles and other important components of a wildland ecosystem. They are unlikely initially to include all the components that allow ecosystems to react to rare disturbance, or climatic, events. Fortunately, relatively functional ecosystems become more diverse over time through natural processes. We should assist this autogenic development by incorporating more functional diversity and functional redundancy than might be suggested by current assembly rules.

Self-design

Self-design introduces as many species and functional groups as appropriate for the particular ecosystem and then allows self-design to sort the species and communities (Mitsch & Cronk, 1992). In response to change, natural systems shift, substitute species, reorganize food chains, adapt as individual species, and ultimately design a system that is adapted to the new environment (Mitsch & Jørgensen, 1989). The ability of ecosystems to organize themselves is one of the fundamental tenants of ecological engineering. Here management is viewed as the 'choice generator' and 'facilitator of matching environments with ecosystems'

(Mitsch & Jørgensen, 1989). It was suggested that ecosystem managers should cultivate the capacity of natural systems to organize themselves rather than trying to control them (Hollick, 1993).

Although the term 'self-design' is not widely known among wildland managers, it is often used, to varying degrees, during wildland repair efforts. Wildland managers often plant many species, when the most effective seeding mixture is unknown. Adding numerous species is a form of insurance against complete failure due to poor or unusual climatic events. Self-design was recommended as the most effective way of developing and maintaining low-input wetlands, since successional processes determine the outcome (Mitsch & Cronk, 1992). In practice, it often begins with dominance by undesirable species, but this is temporary if proper hydrologic conditions are present. Although selective weeding may be necessary in the beginning, ultimately the system develops on its own (i.e., self-designing). The use of nonnative species can increase the benefits of self-organization. When added to well-adapted, diverse ecosystems, nonnative species usually do not dominate the system (Odum, 1989). However, these introductions may create major problems in highly altered environments and islands.

Self-organized systems are most appropriate where the primary objective is a functional ecosystem rather than some predefined composition and/or structure. The abiotic environment determines the direction taken by self-organized systems. For example, we determine the type of wetland by manipulating the hydrology of the site. This approach is effective in creating wetlands that perform specific functions. Self-organized wetlands efficiently filter sewage effluent, remove excessive nutrient loads from runoff water, reduce the acidity of acid mine drainage, and reduce downstream sediment loads (Mitsch, Reeder & Klarer, 1989; Mitsch, 1992; Mitsch & Cronk, 1992; Flanagan, Mitsch & Beach, 1994; Reddy & Gale, 1994; Weiher & Keddy, 1995). However, when land-use objectives require specific products (e.g., wildlife habitat, quality forage, timber, or certain plant species) from the repaired ecosystem, the less predictable outcome of self-design is undesirable. We fine-tune vegetation change by altering the abiotic environment (e.g., nutrients, water, or microenvironment) or with active vegetation manipulation practices (e.g., fire, herbicides, or mechanical plant treatments). Altering the abiotic environment creates self-maintaining systems, but active vegetation manipulation is more rapid and the results are usually better understood.

Which plant part should be planted?

After deciding which species and species mixtures to use, we must determine which form of the plant to use. Should we plant seed, seedlings, or some other part? What factors should we consider when deciding whether to plant seed, seedlings, or plant parts? The cost, effectiveness, availability of plant materials, equipment availability, soil stabilization rate, time frame for achieving repair objectives, and esthetic appearance are important when deciding to plant seed, seedlings, or plant parts (Table 5.1). However, the relative importance of each of these factors is highly site and situation specific.

Seed

Seed are a convenient plant material for repair activities. Compared with other plant parts, they are easy to produce, collect, clean, store, transport, mix, drill, or broadcast. Under the right conditions, they establish relatively easily. Most seed establish well if growing conditions are relatively good and predictable. Developing seedlings are most vulnerable during seedling establishment where the germinating seeds are exposed to predation or desiccation. Other seed, with dormancy mechanisms or long-lived seed banks, are remarkably well suited to waiting years for suitable conditions. Seeding is by far the most common technique for establishing herbaceous plants and is widely used to establish woody plants.

Direct seeding has several advantages over transplanting seedlings or planting parts of plants. Seeds are relatively inexpensive and versatile. Suitable seed are available commercially for many species or collected from locally growing plants. Direct seeding is almost universally used to establish grasses. Their abundant seed production, and a growth form that simplifies mechanical seed harvest, contribute to the popularity of seeding.

Direct seeding tree seed has variable results and the probability of failure is sometimes high (Harmer & Kerr, 1995). For this reason, transplanting seedlings is considered superior to direct seeding for establishing trees (Stevens, Thompson & Gosling, 1990). However, direct seeding is more practical than transplanting on difficult sites (e.g., some reclamation sites, steep slopes, and areas with difficult access). Direct seeding has greater potential for increasing species and genetic variability. Vegetation arising from direct seeding usually has a more natural appearance than that created by transplanted seedlings (Harmer & Kerr, 1995). Although some species are easily established from transplants, it may be less expensive to allow the 'seed of many species to take "pot-luck" among the spatial mosaic of favorable and unfavorable areas than it is to introduce nursery plants' (Packham *et al.*, 1995). This approach is a form of self-design. Selecting woodland seed in England is improved by considering the following: (1) large-seeded, hardy species are better suited to direct seeding than are species with delicate or expensive seed; (2) good quality seed are always preferable; and (3) sufficient seed of each species should be planted (Harmer & Kerr, 1995).

SEED QUALITY

It is important to plant high quality seed, since seed quality varies widely. Seed lots contain varying amounts of inert material, undesirable mixtures, weed seeds, and immature or injured seeds that will not grow. The seed label is an important source of information and a legal requirement in many countries. Labels are not usually required for seed sold locally by farmers, but are more often required when sold by commercial seed dealers. Labels state the kind and variety of the pure seed. It gives the percentage of pure seed, hard seed (seed that fails to germinate during a short germination period), other crop seeds, weed seeds, and inert matter. Percentage germination, excluding hard seed, is on the label. Since seed quality declines with time it is important to check the testing date on the label.

The best seeding rate for a particular situation is influenced by seedbed conditions and the equipment used in planting. Seeding rates are based on Pure Live Seed (PLS) per unit area, where PLS is the percentage of total seed that is alive. It is determined by:

$$PLS = \left[\frac{(\text{Germination \% + Hard Seed \%}) \times \text{Purity \%}}{100} \right] \quad (5.1)$$

Purity is used to reduce the total weight of seed due to trash, straw, or weed seed. Planting weed seed might reduce the establishment of desired species or it might start an extremely noxious weed into a new area, so quality seed lots have fewer weed seed. Hard seed, typically found in legumes, are live, viable seed that do not germinate during a short (10–30 days) germination test. It may germinate later when planted in the soil. Seed are sold in bulk or rates based on PLS, so you must compare among alternative seed lots based on PLS. The PLS price rather than the bulk price determines which of the seed lots is less expensive. The PLS price of seed (Cost$_{PLS}$) is determined by:

$$\text{Cost}_{PLS} = \left[\frac{(\text{Bulk Price}/\text{kg}) \times 100}{\text{PLS}} \right] \qquad (5.2)$$

where Bulk Price/kg is the cost per unadjusted weight.

Seed is the most important means of establishing new plants in most wildlands. However, obtaining sufficient amounts of high quality seed is a serious and continuing problem in many areas. Grass and some forb seeds are grown and harvested with standard agronomic equipment and practices, but the production of shrub seed often requires innovative techniques and hand collection (Monsen, 1985). Commercially grown seed are more readily available, but wildland applications require more ecotypes than seed producers are able to grow. Thus, for many wildland applications, much of the seed will be harvested from wildland sources. The uncontrolled conditions inherent in wildland seed collection can create additional seed quality problems (McArthur, 1988). However, it is usually worth the additional effort to obtain high quality seed.

COLLECTING SEED

Grass seed can be harvested with large or small machinery, bare hands, or small hand-held devices. Hand collection is necessary for many tree and shrub species that cannot be mechanically harvested. Hand harvesting woody plant seed is often practical because fewer seed are required. We should try to collect the seed fully developed. However, strong winds or precipitation may cause the seed to fall to the ground before being collected. The seed need not be completely clean and free of trash, but we should know the purity and germination percentage.

Viable, uninjured seed of two-thirds of American tree species fail to germinate after being processed and tested for germination (Klugman, Stein & Schmitt, 1974). These seed are probably dormant and require specific treatments before they germinate. Moisture, temperature, or light treatments, alone or in some combination, remove most dormancy problems (Klugman *et al.*, 1974). Alternatively, they may require chemical or mechanical seed treatments. Specific knowledge of the dormancy mechanism will help determine which treatments can increase early germination. Untreated seed may germinate in subsequent years.

Several recommendations insure an adequate representation of the local gene pool, when collecting woody plants (Weber, 1986; Barnett & Baker, 1991; Packham *et al.*, 1995). Collect seed from 15 to 25 individuals spaced at least 100 meters apart. Select seed from healthy, unstressed plants. Collect from well-formed, dominant trees. Avoid individuals that are isolated from others of the same species. Restrict harvest to mature seed from ripened fruits. Harvest fruits from all parts of the canopy (top, sides, and bottom) since these parts may are often pollinated from different sources and at different times. Finally, collect throughout a species' normal habitat. In many species, the seed of adjacent plants of the same species mature at different times.

STORING SEED

Seed moisture content, storage temperature, and relative humidity affect the viability of stored seed. Seeds of most terrestrial plants should be stored with moisture contents between 5% and 14%. Generally, starchy seed should be dried to a moisture content of less than 14% and oily seed should be dried to less than 11% moisture (Harrington, 1972; Harrington, 1973). Seeds drier than 5% may be damaged because their cell walls break down and their enzymes become inactive (Apfelbaum *et al.*, 1997). Wetter seeds (> 14%) are more susceptible to bacterial and fungal damage. Above 30% moisture, nondormant seeds may begin to germinate.

Special collection and processing techniques are required for seed from pulpy fruits. The seed should be extracted soon after collection and dried slowly to avoid seed damage. These seed must be stored properly, if not planted soon. Seed remain viable for one to ten years, depending on the species. They are best if used within one year. Most leguminous seeds store well in normal conditions, if properly cleaned, dried, and protected. The seed of other species retain viability for only a few days, months, or at most a year following harvest (Hartmann *et al.*, 1997). Many spring-ripening, temperate-zone trees (*Populus, Acer, Salix, Ulmus*) produce seed that fall to the ground and germinate quickly, thus seed longevity is often very short.

Seed are safely stored in several ways. Seed store well in unsealed containers under controlled temperature and relative humidity conditions. They are effectively stored in well-sealed containers with silica gel desiccant and refrigerated. Polyethylene bags and plastic buckets with sealed lids are effective storage containers. Another effective storage technique is to put the seed in paper or burlap bags and hang them in cool, dry environments. Some seed can be frozen in sealed containers if the moisture content is below 14% (Apfelbaum *et al.*, 1997). It is important to keep seed storage containers off damp floors and keep rodents away from them. Although some impurities do not damage seeds, insect eggs, dirt, or material containing fungus spores can create many problems. To kill insects and fungi in seed containers before storage, place insecticides and seed in a sealed container for 24 hours, then aerate the seed before storage (Ffolliott *et al.*, 1994).

There are three distinct patterns of seed storage behavior (orthodox, recalcitrant, and intermediate) (Murdoch & Ellis, 1992). Orthodox seed (most grasses, forbs, and crop species) are dried without damage. Over a wide range of conditions, longevity increases as seed moisture content and seed temperature decrease (Roberts, 1973). Longevity increases with desiccation to about −350 MPa (Ellis, Hong & Roberts, 1989). These seed should be uniformly dried before storage. Polyethylene bags are good containers because they are impermeable to water, but allow necessary exchanges of oxygen and carbon dioxide (Smith, 1986). This prevents excessive moisture buildup, but still allows the gas exchanges necessary for embryo respiration and continued seed survival.

Recalcitrant seed do not survive desiccation (Roberts, 1973). They are killed at water potentials more severe than −1.5 to −5.0 MPa, or roughly equivalent to the permanent wilting point of many growing tissues. Recalcitrant seed storage behavior occurs in several large seeded woody perennials, such as cocoa (*Theobroma cacao*) and rubber (*Hevea braziliensis*) and in tropical fruits such as avocado (*Persea americana*) and mango (*Mangifera indica*). Temperate timber species such as oak (*Quercus* spp.) and chestnut (*Castanea* spp.) also exhibit recalcitrant seed storage patterns (Murdoch & Ellis, 1992).

An intermediate seed storage pattern occurs in coffee (*Coffea arabica*), papaya (*Carica papaya*), and oil palm (*Elaeis guineensis*) (Ellis, Hong & Roberts, 1990). These seeds are injured at low temperatures when the seed are dried to −90 to −250 MPa (Murdoch & Ellis, 1992).

Some temperate trees, such as beech (*Fagus sylvatica*), may have this type of seed storage behavior (Gosling, 1991).

HAY MULCH AS SOURCE OF SEED

Spreading hay mulch, containing seed, is effective on difficult sites such as gullies, dams, spillways, waterways, dunes, and sand blowouts (Vallentine, 1989). Since hay mulch includes seed from species that are not commercially available, it is often used to plant native prairie species on old fields. Hay mulch applications have the advantage of adding many species and include mulch that can improve establishment. It also has several disadvantages. A single cutting of hay seldom includes all the possible species from an area. Summer harvests miss many spring and fall flowering species. Labor requirements for harvesting and spreading the hay may make this technique too expensive for all but the highest priority sites. Seeding rates are difficult to determine. Hay should be harvested when seed are nearly ripe, but before they shatter and fall to the ground. At least 2000 kg of hay ha^{-1} should be used, but some sites require twice as much. The amount of hay applied has little relationship to the amount of seed applied. Hay mulch applications are most effective when applied on recently prepared seedbeds.

FLUFFY GRASS SEED OR BARE CARYOPSIS

Seed from genera such as *Andropogon, Sorghastrum, Heteropogon, Bothriochloa,* and *Schizachrium* (Figure 5.1) are available as either fluffy seed (with hulls and awns attached) or as bare caryopses. Although almost all the planting of these genera is with fluffy seed, new cleaning processes allow processors to sell bare caryopses. These caryopses are concentrated, so the price per kilogram is higher. This not only reduces seed bulk (10 to 50 times) and transportation costs, it allows the seed to flow more smoothly through seeding equipment. Fluffy seed do not flow easily and require special seeding equipment for their use. In contrast, bare caryopses flow so easily that seeding rates are difficult to set accurately. Bare caryopses are beneficial because they germinate more rapidly. Without precision equipment, the actual planting rate could be unexpectedly high. Hulled buffalograss (*Buchloë dactyloides*), bermudagrass (Burton & Andrew, 1948), and the bare caryopsis of fluffy seed

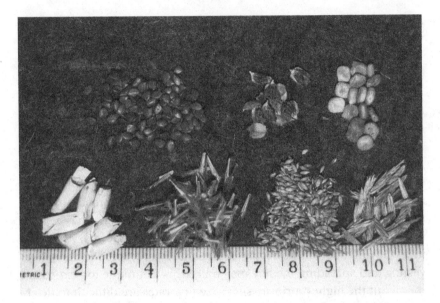

Figure 5.1. The physical characteristics of seed vary greatly and those differences influence the type of equipment necessary for planting. The top row contains forb seed, from left to right: *Desmanthus illinoensis*, *Engelmannia pinnatifida*, and *Lupinus texensis*. Grass seed are in the bottom row, from left to right: *Tripsacum dactyloides*, *Schizachyrium scoparium*, *Panicum coloratum*, and *Bouteloua curtipendula*. Fluffy grass seed (e.g., *Schizachyrium*) have awns or hairs that prevent the seed from flowing through seeding equipment, so they require special equipment. Common seed preparation procedures do not clean all species to bare seeds. For example, the spikelets of some grasses (e.g., *Bouteloua*) and the flower heads of some forbs (*Engelmannia*) are planted. Both the spikelets and flower heads contain many seeds.

germinate more rapidly and are effective when rapid establishment is desired. However, in marginal and variable environments delayed germination may insure against total seeding failure. If dehulled seed germinate rapidly and later die, the unhulled seed might provide another opportunity for establishment.

SEED FROM TOPSOIL ADDITIONS

Topsoil and its accompanying seed bank should be spread on repair sites to improve the soil and add species. The benefits of additional soil are obvious, but the addition of seed and their associated microorganisms (Perry *et al.*, 1989) provide additional benefits. Topsoil additions are relatively common in wetlands, but its cost makes it impractical for most wildland circumstances. Several problems occur when using topsoil for seed. Not only is the amount of seed unknown, the species composition may contain surprises. Species establishment after topsoil additions may differ from the species on the donor site. Noxious weeds, that were not obvious on the donor site, can dominate recipient sites. It is also quite possible that soil removal will create new problems on the donor site. The most desired seed are not always available from topsoil additions. For example, forest topsoils are usually dominated by herbaceous species and early successional trees (Wade, 1989), which is not the objective of most projects.

Whole plants

Transplanting whole plants is the only viable method for establishing some species (Munshower, 1994). For example, planting adapted, drought-tolerant shrubs is the only viable option in many arid environments (Van Epps & McKell, 1980). Direct seeding is less effective because the seed of some long-lived woody plants germinate infrequently and their seedlings grow slowly. While these traits make the plant better adapted to establish during a particular set of climatic conditions, they are poorly suited to direct seeding in wildland ecosystems. Placing whole plants into environments with short, relatively unpredictable environments (such as arid and semiarid environments) is effective because it bypasses the high-risk germination and seedling phases. In difficult environments, transplanting seedlings is more reliable than direct seeding. Transplants start faster and take advantage of short growing seasons. Although the cost per plant is greater, this 'jump-start' increases establishment. Direct seeding is more practical where growing seasons are more predictable and hospitable. Transplanting whole plants has utility for establishing species that (1) usually reproduce vegetatively; (2) have poor germination; and (3) have seedlings with low seedling vigor (DePuit, 1988a). Wildings, bare-root seedlings, and container-grown seedlings each have unique transplanting considerations.

WILDINGS

Wildings are removed from natural settings and transplanted at repair sites (Munshower, 1994). This is successful for trees, shrubs, and herbaceous plants, but is more common for woody species. Wildings are less frequently used than direct seeding or transplanting, because of costs and low wildland survival rates. Transplanted switchgrass plants are most effective at reducing channel erosion. While this species has strong seedlings that establish rapidly, the erosive and depositional forces in channels kill seedlings. Thus, moving nearby switchgrass plants into gully channels is a practical approach.

BARE-ROOT STOCK

Bare-root seedlings should be grown for 8 to 10 months (or more), in outdoor nurseries, before removing them from the soil (lifting) for transplanting. After lifting, the soil should be removed from their roots, and the roots and tops pruned (Figure 5.2). Bare-root seedlings can be tied into bundles for storage and placed in wet paper, cardboard or plastic containers for transport. They must be protected from desiccation during transport with moss, leaves, or prepared mixtures over their roots. Compared with container stock, they are usually hardier, older, easier to transport, less expensive, and do not become root-bound. Bare-root seedlings establish as readily as containerized stock under good conditions. However, containerized stock establishes more readily under dry or otherwise harsh conditions (Vallentine, 1989; Barnett, 1991; Brissette, Barnett & Landis, 1991).

CONTAINER-GROWN STOCK

Containerized planting stock is grown in greenhouses or outdoor facilities in either multiple unit packages or individual containers with a growth medium of soil, vermiculite, or peat moss. Containers are available in many sizes and shapes (Figure 5.3). The peat, paper, cloth, cardboard, or other decomposable containers can be planted with the seedlings, but plastic and metal containers must be removed from the seedling's root plug before planting. Many reusable plastic containers have vertical ribbing (root trainer systems) down the inside that reduces root spiraling in the seedlings. Reusing these containers will offset the higher initial cost. Container-grown stock is widely used in temperate forests, arid and semiarid ecosystems, and wetlands, but is also common in many other areas.

Reusable, root trainer containers are widely used in forests because

Figure 5.2. Bare-root (right) and container-grown (left) seedlings ready for transplanting.

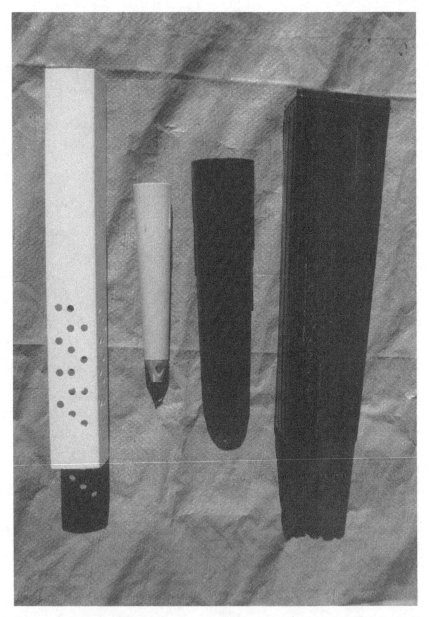

Figure 5.3. Seedlings are grown for transplanting in paper containers that are planted with the seedlings (left) or in reusable plastic containers. Both types are available in many sizes, since transplanted seedlings with larger root volumes are more expensive but have greater establishment rates in difficult environments.

of their relatively low cost and effectiveness. In the southeastern US, plantable loblolly pine (*Pinus taeda*) and slash pine (*Pinus elliotti*) are grown in 12 to 14 weeks, and longleaf (*Pinus palustris*) in about 16 weeks using containers (Brissette *et al.*, 1991). This rapid production has significant advantages for wood production systems. When necessary, replanting of spring-planted areas can be conducted during fall of the same year, saving a full year over bare-root methods (Brissette *et al.*, 1991). When planted in early spring, on high quality bottomland, bare-root loblolly pine seedlings performed as well as container-grown seedlings (Barnett, 1991). However, container-grown stock outperformed bare-root stock on lower quality sites and when planting was conducted later in the spring. Container-grown seedlings outperformed bare-root stock on adverse sites and were the only way to establish trees on severely damaged sites.

Container-grown seedlings are the most reliable method for establishing woody plant seedlings in arid and semiarid ecosystems (Vallentine, 1989; Munshower, 1994). While bare-root seedlings of many shrub and forb species require 1 to 2 years growth before transplanting into arid or semiarid rangelands, containerized seedlings may be transplanted after only 12 weeks (Vallentine, 1989). Long, narrow (40 × 5 cm) paper containers (plant bands) are increasingly being used to grow shrub seedlings destined for desert planting. These containers develop deep root systems, particularly when watered from below instead of sprinkling from above. These plant bands are more effective if not removed before planting (Felker, Wiesman & Smith, 1988). Plant bands and reusable plastic containers were used in a comparison of *Prosopis alba* and *Leucaena leucocephala* seedling establishment in semiarid south Texas (Felker *et al.*, 1988). In a dry year the plant band seedlings had a higher survival rate than seedlings grown in plastic containers, but no differences were observed in a wet year (Felker *et al.*, 1988).

Container-grown plants tolerate competition and harsh environmental conditions more readily than seeded or bare-root transplants. In southern California, where annual weeds prevent natural recovery of coastal sage scrub, recovery was made possible with container-grown stock (Eliason & Allen, 1997). In the Mohave Desert, the establishment rate of transplants was increased significantly when seedlings were grown in 76 cm tall sections of plastic pipe (Holden & Miller, 1993). Additional root depth and volume provide significant advantages that can make the difference between success and failure in arid environments.

Containerized stock dominates wetland applications because direct seeding has been too slow and/or unpredictable. Bare-root stock is largely unavailable for herbaceous plants. Woody bare-root stock works well on the best sites, but container-grown stock survives on sites that are too harsh for bare-root seedlings (Clewell & Lea, 1990). Container-grown stock can be planted later in the growing season than bare-root seedlings (Clewell & Lea, 1990). However, container-grown seedlings cost several times that of bare-root seedlings, are more difficult to plant, and should be limited to unsuitable sites for bare-root stock because of economic considerations (Clewell & Lea, 1990).

Compared with bare-root stock, there are several advantages to growing and planting seedlings raised in containers:

1. Seedlings are started at any time in the greenhouse.
2. Rare seed are most efficiently used because of the higher survival rate compared with direct seeding or bare-root transplants.
3. Rapid growth reduces the time required for quality seedlings.
4. Containerized stock is less susceptible to injury during transport.
5. Containerized stock has fewer storage problems after transport.
6. Transplanting time is longer for containerized stock.
7. Containerized stock has intact roots that are well developed when planted.
8. Containerized stock is less susceptible to shock after planting.
9. Growth in a container insures faster root development and larger and healthier plants after one or two growing seasons than comparably sized bare-root stock or wildings (Munshower, 1994).
10. Field survival is higher under difficult conditions (Vallentine, 1989; Barnett, 1991; Brissette *et al.*, 1991).

SOD

Transplanting shallow layers of soil with plants rapidly establishes turf in yards and golf courses, but the expense of this method has limited its application on wildlands. Disturbed sites are rapidly stabilized with sod, but it is limited to high priority sites subject to severe erosion or sites requiring rapid stabilization (Munshower, 1994). It is used to

Figure 5.4. Dormant willow stems (1–2 m long) were placed vertically into active dunes near Yulin, Shaanxi Province, People's Republic of China (with only 4–10 cm remaining above the soil surface). They established rapidly and began to capture sand and other wind-blown materials while greatly restricting dune movement.

transfer prairie sod from future construction sites to a protected site. Sodding is a highly effective way to transfer many plant species, associated soil organisms, propagules, and some nutrients, and organic matter.

Plant parts

Asexual reproduction from stolons, rhizomes, root sections, stem sections, cactus cladophylls, or other parts is common. New plants should be propagated, for transplanting, in the greenhouse or field site with adventitious buds. Sprigging stem and stolon sections is the only reliable propagation method for coastal bermudagrass (variety of *Cynodon dactylon*) since it seldom produces viable seed (Burton & Hanna, 1985). Transplanting tree stem-sections is relatively inexpensive and adapted to a large numbers of trees or shrubs. Willows (*Salix* spp.) are propagated in riparian ecosystems from cuttings (Monsen, 1983) and are used in many countries to stabilize active dune sand (Figure 5.4).

6

Site preparation and seedbed management

Introduction

Selectivly manipulating the seedbed environment is a powerful tool that influences the direction of repair efforts. We selectively modify the availability of safe sites to influence the direction of change. Well-designed site preparation strategies increase certain species by creating safe sites that favor their establishment over unwanted species. Some site preparation techniques focus on improving equipment access to sites with heavy debris or vegetation.

Well-designed site preparation and seedbed management strategies consider the long-term implications of vegetative development on microenvironmental parameters. We can manipulate site and propagule availability by (1) doing nothing (unassisted natural recovery); (2) modifying seedbed conditions and relying on natural seed dispersal mechanisms (assisted natural recovery); or (3) selectively modifying seedbed conditions and adding carefully selected species mixtures (artificially induced recovery).

Unassisted natural recovery

Unassisted natural recovery is a passive strategy that requires neither seedbed preparation nor planting (Table 6.1). This strategy is attractive because it requires no management input. The direction and rate of natural recovery processes are restricted by (1) insufficient immigration of desired seed; (2) dominance by undesirable plants; and (3) excessive herbivore damage to developing plants. Careful evaluation of these potential interactions contributes to the development of appropriate recovery strategies. Chapter 4 contains additional discussions of natural recovery strategies.

Unassisted natural recovery operates slowly, with unpredictable results (Harmer & Kerr, 1995). It is unsatisfactory if repair objectives require the establishment of particular species groups within a particular time span. Colonization requires years to develop, particularly if it does not begin soon after disturbance. For example, a study of 47 derelict sites in central and west England found that only 30% of sampled quadrats were colonized, by one or more woody plants, within 10 years, and this changed little for another 25 years (Harmer & Kerr, 1995). Unassisted natural recovery was most effective when used to augment populations rather than to create certain species assemblages.

Since successional changes are driven, in part, by differential species availability, landscape configuration is a critical factor in the effectiveness of unassisted natural recovery strategies (Whisenant, 1993). The arrangement of landscape components partially defines each site's role as propagule donor or recipient. Unassisted natural recovery is greatly inhibited in large damaged landscapes without patches of undamaged vegetation and soils. For example, the large-scale conversion of Arizona deserts to cultivated fields left little of the original desert (Jackson, 1992). Most of the 220 000 ha Santa Cruz Valley was cleared for farming. About half that area is now abandoned. Secondary salinization, no natural seed sources, and the arid environment limit recovery on these fields (Jackson, 1992). Wind-blown species (*Isocoma tenuisecta* and *Baccharis sarothroides*) and livestock-dispersed species (*Prosopis velutina*) now dominate much of the area. However, the original, long-lived, dominants (*Larrea tridentata* and *Atriplex polycarpa*) have heavy, poorly dispersed seed (Jackson, 1992). Traditional artificial seed strategies are unlikely to solve these large-scale problems, due to the great costs involved.

Unassisted natural recovery usually creates landscapes dominated by species from the surrounding sites. This dominance may last several decades. For example, in England, unassisted natural recovery often produces communities dominated by species with light wind-blown seed, such as willow (*Salix* spp.) and birch (*Betula* spp.) or species with

Table 6.1. *Comparison of unassisted natural recovery, assisted natural recovery, and artificially induced recovery strategies*

Strategy	Description	Advantages	Disadvantages
Unassisted natural recovery	No seedbed modification Reliance on natural seed dispersal mechanisms	Low establishment costs Low labor or equipment costs Little soil disturbance Does not depend on the availability of commercial seed or nursery-grown seedlings Few disease or insect problems	May be slow when seedcrops are low Reduced control over initial density, pattern, spacing or timing Not harvesting seed trees reduces (or delays) income May require precommercial thinning of trees Competing vegetation may reduce establishment success Species changes are difficult to direct
Assisted natural recovery	Selective seedbed modification Reliance on natural seed dispersal mechanisms	Moderate establishment costs Moderate labor or equipment requirements Does not depend on the availability of commercial seed or nursery-grown seedlings Few disease or insect problems Control of competing vegetation increases success Selective seedbed modification allows some directed species changes	May be slow when seedcrops are low Moderate to high soil disturbance with variable time required for return of protective soil cover Delayed establishment and soil disturbance increase erosion More expensive than unassisted natural recovery Reduced control over initial density, pattern, spacing or timing Not harvesting seed trees reduces (or delays) income May require precommercial thinning of trees
Artificially induced recovery	Selective seedbed modification	Good control over spacing, pattern and initial density	Establishment costs may be high Labor and equipment requirements may be high

Artificial addition of selected plant materials	Control of competing vegetation increases success	Insect and disease problems may be serious with some species
	Not limited to or dependent on natural seed crops	Poorly chosen species mixtures may not achieve objectives or persist
	Permits directed species changes	

Source: Adapted and modified from Barnett & Baker (1991).

bird-dispersed seed (Harmer & Kerr, 1995). This vegetation is typically impoverished, with little commercial value. While unassisted natural recovery creates fully vegetated sites, humans have little influence on its composition unless we selectively influence seed immigration. Assisted natural recovery or artificially induced recovery strategies can direct the direction and rate of recovery.

Some sort of vegetation management (e.g., herbicides) is usually necessary on weed-dominated sites (Harmer & Kerr, 1995). The most severe weed problems typically occur on fertile soils. Thus, natural recovery is often more successful on infertile soils with little competing vegetation (Harmer & Kerr, 1995). As an example, in Ontario, Canada, unassisted natural recovery effectively established diverse vegetation on vertical cliff faces of abandoned limestone quarries (Ursic, Kenkel & Larson, 1997). Site age and the density of trees adjacent to the quarry walls had the greatest effect on the vegetation composition and abundance.

Assisted natural recovery

Assisted natural recovery uses locally produced seeds that arrive on the site through natural methods (Table 6.1). This approach is widely used in rangelands and forests because it is a practical, low-cost alternative to planting, where adequate seed sources are available (Barnett & Baker, 1991). We control the rate and direction of recovery with seedbeds that benefit some species over others or by influencing which seed arrive on site through natural means. Chapter 4 described examples of differential establishment from various seedbeds (see sections on differential site availability and safe sites).

Forest clear-cuts surrounded by seed-producing plants recover without artificial seeding. Assisted natural recovery relies on a few seed trees or groups of trees (shelterwood) within the clear-cut area. In the southeast US, seed-tree methods reliably reestablish longleaf pine forests following logging (Stoddard & Stoddard, 1987). Seed-tree strategies have several advantages: (1) less expense due to reduced labor and equipment costs; (2) greater esthetic appeal; (3) improved wildlife habitat; and (4) since seed trees add seed for several years, this approach is less affected by a single poor seed crop than a direct-seeding operation (Cubbage, Gunter & Olson, 1991). However, this approach to assisted natural recovery has several disadvantages in forests, including: (1) there may be too many seedlings; (2) seedlings tend to be too sparse or irregularly spaced; (3) costly weed control may be necessary; (4) it typically produces less wood volume than even-aged management; (5) scattered trees (particularly conifers) often blow down during wind storms; (6) seed trees may be difficult to salvage without damaging new trees; and (7) this strategy precludes the introduction of new germplasm (Stoddard & Stoddard, 1987; Cubbage *et al.*, 1991). For these reasons, the seed-tree method is less effective in northern pine forests (Stoddard & Stoddard, 1987). A study of 30 sites in southern England revealed that the presence of seed trees does not insure their progeny will become established (Harmer & Kerr, 1995) because of: (1) too few seed produced; (2) excessive seed predation; (3) competitive exclusion by other vegetation; or (4) browsing damage.

The minimum number of seed trees depends on the seed production of the species and individual tree. In most cases, at least 10 to 25 seed trees per hectare should remain (Stoddard & Stoddard, 1987). Although seed dispersed by mammals or birds may move great distances (see Chapter 4), the great majority of wind-dispersed seed will be deposited within a few tree heights of the seed tree (Greene & Johnson, 1996). Seed trees should remain for at least 5 years or until adequate reproduction occurs. If an adequate seed source is available, the most important elements for successful natural recovery are effective site preparation and control of competing vegetation (Barnett & Baker, 1991). The most effective type of site preparation depends on the type of trees desired. The large seed of longleaf, slash, and ponderosa (*Pinus ponderosa*) pine require an exposed mineral soil seedbed. Some species require exposed mineral seedbeds only when seed production is low (Barnett & Baker, 1991). Generally, for the southern pine species, the soil disturbance during logging exposes sufficient mineral soil to stimulate adequate recovery.

This same general strategy is appropriate for damaged grasslands and shrublands with remnants of desired vegetation. It reduces expenses, but requires time for natural recovery. This approach is the basis of repair strategies aimed at management of the ecosystem's natural producers rather than the introduction of additional plants.

Artificially induced recovery

Artificially induced recovery strategies are most appropriate where site degradation is too great for natural recovery or where more rapid results are desired (Table 6.1). Artificially induced recovery strategies modify safe site and species availability by adding seed (or plants) and actively manipulating the frequency of safe sites. We have the most influence on the pace and direction of change when we control seedbed conditions and species availability. When properly designed and conducted, artificially induced recovery initiates, directs, and accelerates natural processes. Knowledge of why seedings fail, how to overcome this problem, and how to plant seed, plant parts, or whole plants is essential to the use of artificially induced recovery strategies.

Why do seedings fail?

Seedings fail due to problems encountered during the germination, emergence, and establishment phases (Decker & Taylor, 1985; Vallentine, 1989; Harmer & Kerr, 1995). Improper techniques caused most of the pine seeding failures in the southeast US (Barnett & Baker, 1991). Most of the failures were associated with seeding unsuitable sites, seeding at the wrong time, inadequate site preparation, poor-quality seed, and too few seed. In west Texas, successful grass seedings were related to the degree of seedbed preparation, rainfall, and temperature (Stuth & Dahl, 1974). Most of the serious problems encountered during woodland seedings, in England, were associated with predation by birds and mice, low germination, and weed competition (Harmer & Kerr, 1995).

Wildland seeding involves considerable uncertainty. Much of the uncertainty is due to unpredictable weather. While we have no influence over the weather, certain seedbed management technologies ameliorate the detrimental effects of harsh environmental conditions. Effective site preparation and planting practices reduce failures (Coppin & Stiles, 1995). Developing practical strategies requires that we consider the probability of success and the consequences of failure for all possible alternatives.

GERMINATION AND EMERGENCE
Many of the same problems affect germination and emergence of seeded plants. Once planted, seed are subject to potentially harmful moisture and temperature conditions. Extreme temperatures reduce germination. Soil moisture may be too low or sufficiently high to restrict soil oxygen availability in some soils. Low-quality seed or poor seedbed conditions cause most germination problems. Low-quality seed, poor planting techniques, predation, and surface soil crusts cause many emergence problems. Having the seed tested, using good seed, and using them promptly after harvest reduces problems associated with low-quality seed. Seed storage conditions affect seed quality. Large seed survive storage longer than small seed. Cool, dry conditions increase storage life, as long as insects or disease do not damage the seed.

Good seed-to-soil contact provides a more reliable water supply to the seed. Seed planted in excessive amounts of litter or in large air pockets are less likely to become established. The soil should be firmed above and around the seed to reduce large air pockets. Excessive amounts of organic materials cause similar problems by preventing seed/soil contact directly or by preventing closure of the drill slit (Marshall & Naylor, 1984). Seeding and seedbed preparation techniques that reduce the amount of litter increase seedling emergence by increasing seed-to-soil contact and reducing evaporation from drill slits (Marshall & Naylor, 1984).

Seeding too deep, soil crusts, desiccation, wind erosion, water erosion, rodent depredation, insect damage, excessive soil salinity, and frost heaving reduce seedling emergence. Birds, rodents, and insects often eat seeds before they germinate, especially seed broadcast over the soil surface without any attempt to cover them. Soil crusting is a major factor contributing to grass seedling mortality (Rubio *et al.*, 1989).

ESTABLISHMENT
Environmental conditions or resource limitations often limit seedling establishment. Plants do not consume 'conditions' (temperature, soil pH, salinity, or soil bulk density), and we manage those limitations with appropriate seedbed preparation techniques, planting dates, and well-adapted species. 'Resources' (light, CO_2, water, and nutrients) are

consumed by growing plants and are either (1) continuously available (not depleted) in limited amounts without storage reserves (i.e., light); or (2) available in limited amounts that are stored or depleted (i.e., nutrients in infertile soils; water in water-limited environments). This distinction, between conditions and resources, is important because it indicates the importance of early weed control. For example, water transpired through weeds is unavailable for seeded species, even after weed removal. Nitrogen taken up by weeds is largely unavailable to the seeded species during the same growing season. Early weed control is preferable to later weed control, particularly if the factor limiting seedling establishment and growth is a resource.

Seedling establishment during either natural or artificial recovery is a function of the number of seeds in safe sites rather than the total number of available seeds (Harper *et al.*, 1965). Well-prepared seedbeds improve environmental conditions and improve resource availability by controlling established weeds. Planting strategies that maximize the abundance of safe sites and accurately place seed may be less expensive because they require less seed and reduce the risk of failure. Therefore, it is particularly important to carefully plan and implement site and seedbed preparation strategies.

Seedbed preparation

Seedbed preparation is the primary concern of most wildland repair activities, since it is the most labor-intensive, is energy consumptive, and often determines success or failure. Strategies that repair soil surface processes also improve seedbed conditions and increase seedling establishment. These treatments repair damaged primary processes (infiltration, runoff, nutrient cycling) critical to establishment and long-term success. Ideal seedbeds are (1) firm below and above seeding depth; (2) composed of thoroughly tilled, friable soil; (3) not cloddy or compacted; (4) devoid of established weeds; (5) without significant seedbank of weedy species; and (6) covered with moderate amounts of mulch or plant residue on surface (Vallentine, 1989). While these conditions benefit most seeded species, other species may require mineral-soil seedbeds, shade, or nurse plants.

Weed-free seedbeds increase grass, forb, shrub, and tree establishment in most wildland ecosystems (Evans *et al.*, 1970; Nelson *et al.*,

176

1970; Stuth & Dahl, 1974; Evans &Young, 1975; Evans &Young, 1978; Roundy & Call, 1988; Vallentine, 1989; Munshower, 1994; Snow & Marrs, 1997). Competition from herbaceous weeds is one of the most common causes of seeding failure and the degeneration of established seedings. Weeds reduce the establishment by depleting limited resources such as nutrients or water and intercept continuously available resources (e.g., light). In both situations, early emerging seedlings gain an immediate and lasting advantage over their later emerging neighbors. This seedbed competition leads to a dominance hierarchy that eliminates the smallest individuals (White & Harper, 1970). Seedling survival is largely determined by position in that hierarchy (Ross & Harper, 1972). Weed control may have lasting effects on community growth and productivity. For example, ten years after planting loblolly pine (*Pinus elliotii*) into a silt loam soil, plants that were seeded into disked strips and furrows still had higher survival rates and were 1.0 to 1.5 m taller than those seeded into grassy seedbeds (Lohrey, 1974).

The objectives of seedbed preparation are to create safe sites for seeded species. Selective disturbances effectively manipulate the type, severity, season, and dispersion of disturbance. This determines safe site availability. Selecting from among the many types of disturbance depends on the kind and amount of existing vegetation and soil factors such as susceptibility to erosion (by eolian or fluvial processes), slope, salinity, stoniness, texture, and depth. Accessibility to seeding equipment, cost limitations, obstructions, and value of resident vegetation also dictate the choice of seedbed preparation methods. Individual seedbed preparation methods have different effects, but they all address the critical problem of weed management. The common categories of manipulating sites and seedbeds are (1) mechanical or manual; (2) chemical; (3) fire; (4) biologic; and (5) the use of mulches (covered in subsequent section).

MECHANICAL AND MANUAL METHODS

Recently harvested forest sites have standing and downed woody debris. This debris is rearranged or removed before additional site preparation activities (burning, disking, or chopping) occur (Lowery & Gjerstad, 1991). This debris is often knocked down, chopped, and crushed before planting (Figure 6.1). Since these activities expose large

Figure 6.1. Roller chopper used for site preparation in east Texas forest. This treatment breaks up the woody debris and roughens the soil surface to facilitate and improve planting success.

areas of mineral soil, they must be conducted with great caution in areas with erodible soils and sloping terrain to avoid accelerated erosion (Lowery & Gjerstad, 1991).

Although hand labor is effective, mechanical seedbed preparation often involves standard agricultural techniques such as plowing, chiseling, disking, or harrowing. Farm or other specialized equipment opens and roughens the soil surface, kills existing vegetation, and facilitates the planting process. Manual site preparation techniques are labor intensive. Farm equipment is capital intensive and difficult to use on steep, rocky slopes. Farm equipment causes additional problems if it packs the soil. Both mechanical and manual methods loosen the soil, reduce soil surface crusts (at least temporarily), direct water into depressions, and reduce wind speed and temperature extremes for developing seedlings. Germination and survival usually increase when the soil surface is cultivated and well prepared before planting. For example, in the heathlands and moorlands of Scotland, tree seedling establishment on disturbed soils was ten times greater than on undisturbed ground (Miles & Kinnaird, 1979). Similar responses occur in many other ecosystems.

While disturbance (cultivation) increases seedling establishment, it also increases erosion risks. Clean seedbeds are effective where wind and water erosion are not serious problems and establishment is not greatly limited by precipitation. However, clean seedbeds are more susceptible to erosion on sandy soils, slopes, or other erosive situations. Soil disturbance (cultivation) is generally not recommended where the soils are loose and slopes exceed 20% (Banerjee, 1990).

Although removing the existing vegetation can accelerate erosion, not removing competing vegetation almost certainly reduces seedling establishment (Banerjee, 1990). Blowing sand, released after vegetation removal, causes additional problems by burying seeds, exposing other seeds, and killing young seedlings. Creating furrows to roughen the soil surface reduces many problems, but complicates seed placement for most equipment. In semiarid and arid wildlands, survival is highest when the seed are placed at the bottom of furrows. Grass seed planted on the ridges between the furrows, or on the south facing sides (in Northern Hemisphere), have higher mortality (Hull, 1970). Where the climate and soils allow long periods of standing water, seedling establishment is greater on the furrow tops.

Compacted surface soils and subsoils are common site preparation problems (Brown *et al.*, 1978; Berry, 1985; Cotts *et al.*, 1991; Davies *et al.*, 1992; Sopper, 1992). Ripping (Figure 6.2) or subsoiling involves pulling a steel shank through soils to break up compacted subsurface layers. The shanks are over 45 cm long and spaced about the same distance apart (Munshower, 1994). Deep ripping is effective at reducing the detrimental impacts of surface and subsurface compaction (Berry, 1985; Ashby, 1997; Bell *et al.*, 1997; Luce, 1997). Ripping or subsoiling to break compacted soils (or subsoil layers) facilitated the restoration of jarrah forest vegetation after bauxite mining in southwestern Australia (Ward, Koch & Ainsworth, 1996). Ripping or subsoiling increased the precipitation use efficiency on semiarid rangelands (Wight & Siddoway, 1972), and helped restore native plants to abandoned roads in western Wyoming (Cotts *et al.*, 1991).

Compaction-related problems of cultivated fields were greatly improved, within 2 to 4 years, after planting to pasture species (*Pennisetum clandestinum*). Surface and subsurface macroporosity increased, aggregate stability increased, and surface crusting diminished (Bell *et al.*, 1997). Final steady-state infiltration rates under well-managed pastures increased four-fold compared with continuously

Figure 6.2. Deep ripping (0.4–0.6 m) this highly compacted substrate in west Texas was necessary before water would move into the soil. Ripping or deep plowing was necessary to get any establishment of planted species in this seedbed.

cropped soils (Bell *et al.*, 1997). Although vegetation did not reduce compaction below 15 cm, soil faunal activity and root penetration increased hydraulic conductivity in the compacted layers.

Plowing improves crusted or compacted surface soils, at least temporarily, and kills or damages any competiting vegetation. Moldboard plows (Figure 6.3) incorporate organic materials and bury weed seed deeply enough to prevent emergence, but have high-energy requirements. Chiseling or disking temporarily break surface crusts and kill shallow-rooted weeds. Harrowing reduces soil clods before drilling.

When to cultivate and how often to cultivate usually depends on previous land use and the resulting vegetation. The aim is to plant weedfree sites at the optimum time with a minimum of additional preparation. Sites with an abundance of weed seeds require additional steps to reduce their effect on the planted seed. Sites with a dense cover of annual and perennial vegetation need frequent cultivation and/or herbicides to remove persistent, problem weeds.

Creating a firm seedbed is an important component of effective seedbed preparation when establishing grasses and forbs (Decker &

Figure 6.3. Moldboard plow used to loosen compacted soil and incorporate leaf litter into an old roadbed.

Taylor, 1985). Firm seedbeds hold water near the surface and make it easier to control seeding depth. The final mechanical operation should leave a seedbed that is loose enough for good water infiltration and firm enough to support seeding equipment and provide good seed-to-soil contact. The most common problem with mechanically prepared seed-beds is a loose, soft seedbed. Firming improves surface-soil water retention long enough for the seedlings to establish. Packing is more effective with adequate soil moisture and is less effective on dry, light-textured soils. Packing with smooth rollers can be very detrimental under wet conditions because the smooth surface is more subject to wind and water erosion (Vallentine, 1989). Rolling to firm loose seed-beds prior to drilling is most effective (Hyder *et al.*, 1961). Rolling after broadcasting to cover seed and firm the soil is effective on freshly plowed seedbeds where compaction above the seed is not excessive (Vallentine, 1989).

The completed seedbed is relatively smooth for drilling and rough for broadcasting. Chiseling (or disking) followed by a harrow and some type of cultipacker produces a good seedbed for drill seeding. Flexible cultipackers are more effective than smooth rollers, because they adjust to uneven terrain and pack the soil more evenly. Chiseling or disking to

Figure 6.4. Disk-chain-diker plows the surface and creates numerous depressions in a single pass. These areas are then broadcast seeded (often aerially) without any seed covering treatment. Photograph courtesy of Harold Wiedemann.

10 cm produces a good seedbed for broadcast seeding (Munshower, 1994).

Disk-chain-dikers have disks welded to the links of a large anchor chain. The chain rotates as it is pulled behind a crawler tractor; creating about 40 000 basins ha^{-1} arranged in a pattern of diamond shaped basins approximately 10 cm deep (Wiedemann & Cross, 1990). Attaching a chain-diker behind the disk-chain improves tillage, land smoothing, and basin formation, in a single pass (Figure 6.4). This equipment requires high horsepower crawler tractors, but is very effective on sites with relatively large amounts of brush or woody debris. These seedbeds are well suited for aerial seeding. Broadcast seeders attached to the crawler tractor or the disk-chain-diker apply seed to the area in one operation. This equipment does not cover the seed. However, the next rainfall event erodes soil to the bottom of each pit and covers the seed.

Land imprinting uses heavy rollers (Figure 6.5) to make an imprint in the soil surface. This provides better control over infiltration, runoff, and erosion (Dixon, 1990). Imprinting is most effective on sites with

Figure 6.5. Land imprinter used to create depressions in the soil surface. Large seed are broadcast in front of the imprinter and mashed into the soil of the depressions. Very small seed are broadcast over the depressions behind the imprinter. Photograph courtesy of Warren Clary.

few competing plants and sandy or loose soil (Clary, 1989). Imprinting is the most effective direct seeding technology in the hot, dry Mohave Desert environment because it concentrates rainfall in the basins formed by the imprinter (Holden & Miller, 1993). Seed are often broadcast in front of the imprinter and pressed firmly into contact with the soil (Anderson, 1981), or broadcast behind the imprinter so that splash erosion covers seed in the depressions. Very small seeds that are buried too deeply in loose soils by land imprinters (Roundy, Keys & Winkel, 1990) are most effectively broadcast behind the imprinter.

CHEMICAL METHODS

There are at least two very different uses of chemicals as seedbed preparation methods: (1) herbicides for weed control; and (2) polyacrylamides to aggregate soil particles and prevent the development of soil crusts. Although often combined with mechanical seedbed preparation, they are also effective when used alone. Chemical weed control has several advantages over mechanical weed control. Herbicides (1) leave firm seedbeds; (2) do not increase erosion; (3) are preferred on rough, rocky sites; (4) are sometimes more selective than mechanical methods; (5) conserve soil moisture; (6) are often less expensive; and (7) are occasionally applied when seeding (Vallentine, 1989). Herbicides have the possible disadvantages of (1) not controlling all weeds; (2) being less useful when the seeding mix contains grasses, forbs and shrubs; and (3) leaving enough litter on the soil to interfere with planting or seedling establishment.

Herbicides improve the establishment of planted species by controlling competing vegetation without damaging planted species. This requires herbicides that are either physiologically selective or selectively applied. Properly used, herbicides with the desired physiological selectivity will control competing vegetation without damaging desired species. Herbicides that damage seeded species are only useful if applied at a time or place that limits damage to the seeded species. Foliar-absorbed herbicides are useful when applied directly to target plants or applied before planting, providing they have no lasting soil activity. Soil- or foliar-active herbicides can maintain a chemical fallow that reduces weed seed and limits soil water losses to weeds. Chemical fallow techniques provide excellent control over wind and water erosion at less cost than mechanical fallows with several tillage operations (Good & Smika, 1978).

Herbicides are increasingly applied for forest site-preparation in the southeastern US. This increase occurred because (1) new herbicides became available; (2) they minimize soil damage; (3) they are less expensive than mechanical methods; (4) tree growth is greater on chemically prepared sites; and (5) control of competing woody vegetation lasts longer (Lowery & Gjerstad, 1991). Fire often follows chemical applications to increase control of competing plants, reduce harvest debris, and facilitate planting operations.

In southeast Washington, sagebrush (*Artemisia* spp.) establishment following wildfires increased on areas receiving spring herbicide applications (Downs, Rickard & Caldwell, 1993). Studies in Nebraska demonstrated the value of herbicides as a weed control tool during the establishment of native prairie vegetation (Martin, Moomaw & Vogel, 1982; Masters, 1995; Masters *et al.*, 1996). Atrazine (6-chloro-N-ethyl-N'-[1-methylethyl]-1,3,5-triazine-2,4-diamine) controls weeds during the establishment of big bluestem and switchgrass (Martin *et al.*, 1982). If atrazine is not available, metolachlor [2-chloro-N-(2-ethyl-6-methylphenyl)-N-(2-methoxy-1-methyethyl)acetamide] is a suitable replacement for controlling most weeds when establishing big bluestem (Masters, 1995). Imidazolinone herbicides [(+)-2-(4-iso-propyl-4-methol-5-oxo-2-imidazolin-2-yl)-5-methylInicotinic acid] effectively controlled weeds during the establishment of big bluestem, switchgrass, little bluestem, blackeyed-susan (*Rudbeckia hirta*), purple prairieclover (*Dalea purpurea*), Illinois bundleflower (*Desmanthus illinoensis*), trailing crownvetch (*Coronilla varia*) and upright prairie coneflower (*Ratibida columnifera*) on cropland (Masters *et al.*, 1996).

While there are numerous additional uses of herbicides during seedbed preparation, their use is too site- and species-specific for any comprehensive description here. No universal recommendations for herbicide use as a site preparation tool are possible because of differences in legal restrictions and species tolerance. Soil and environmental factors will significantly affect herbicide movement and persistence. Consequently, effective chemical site preparation requires a thorough knowledge of herbicide effects on the species and the environment. Local knowledge of herbicides, their fate in the environment, and their selective application for wildland repair activities should be used to guide the development of effective strategies.

Polyacrylamides (PAM) are soil conditioners that reduce crust formation, increase infiltration, and improve seedling emergence (Wallace & Wallace, 1986; Rubio *et al.*, 1989; Rubio *et al.*, 1990; Rubio *et al.*, 1992), even at low concentrations (Wallace & Wallace, 1986). Applying PAMs, as granules or water-solutions, alleviated serious crusting problems. They stabilized the soil by flocculating the finer soil particles and binding or gluing them together. These water-stable aggregates are more resistant to erosion and more conducive to seedling establishment. Grass emergence on hard crusted soils in New Mexico was increased when PAMs were used (Rubio *et al.*, 1992).

BURNING METHODS

Prescribed burning has several potential uses for preparing seedbeds. In forests, fire stimulates natural recovery of some tree species and reduces woody debris that interferes with livestock, wildlife, or planting equipment. In grasslands, fire removes herbaceous litter that interferes with mechanical seedbed preparation and facilitates interseeding into the existing vegetation. When misused, fire accelerates soil erosion and damages property. Fortunately, well-developed prescribed burning techniques allow knowledgeable practitioners to accomplish site-preparation objectives.

The effectiveness of fire in debris removal varies with environmental conditions and the amount and distribution of fuel. Burning conditions favorable for debris removal are more hazardous than burning conditions for maintenance burning. The ignition and consumption of woody debris requires hotter, drier environmental conditions. High-intensity fires can damage some desired species. So, we must carefully assess the potential damage against the expected benefits.

Two types of burning for debris removal occurs in coniferous forests (Van Lear & Waldrop, 1991): burning the debris where it lies or burning it in piles or rows. Burning windrows and piles removes a higher percentage of the wood than broadcast burning. Removal of large debris is most important before machine planting. Humus, topsoil, and mineral soil should remain in place, with minimal disturbance during the creation of piles and windrows.

Prescribed fires stimulate natural recovery of trees requiring bare mineral soil seedbeds. Low-intensity fires in the southeastern US stimulate loblolly pine recruitment (Van Lear & Waldrop, 1991). Pine establishment was stimulated on clear-cut sites surrounded by suitable seed trees. In the southwestern US, ponderosa pine (*Pinus ponderosa*) germination was much higher on burned sites than on sites with deep duff (partially decomposed litter). Seedling survival requires long taproots to prevent desiccation during autumn droughts and to resist frost-heaving damage. Roots of ponderosa pine seedlings remained in heavy duff on unburned sites without developing the deep taproots of seedlings on burned sites (Sackett, Haase & Harrington, 1994).

BIOLOGICAL METHODS

Biological seedbed preparation includes nurse crops, preparatory crops, and woody plants to ameliorate harsh soil and microenvironmental conditions. Although each of these three methods requires two separate plantings, the timing of those plantings is different. Nurse crops and woody plants are usually grown simultaneously with the desired species, but preparatory crops are grown and harvested (or plowed under) prior to planting the final species. Effective repair strategies not only address initial establishment concerns, they initiate autogenic processes that continue to improve seedbeds, and facilitate the long-term recruitment of additional plants (Danin, 1991; Jones *et al.*, 1994; Whisenant, 1995; Whisenant *et al.*, 1995).

Nurse crops, also called companion crops, often help establish improved pastures in humid regions and irrigated pastures. Under these conditions planting nurse crops at or near the time when the perennial species are planted has several advantages, including: (1) it reduces wind and water erosion; (2) there is less weed competition; (3) seedlings are sheltered from wind and severe temperature; and (4) the nurse crop provides forage before the perennial species are fully developed. The competition, from nurse crops, must be controlled (partitioned in time or space) to increase perennial species establishment. Nurse crops delay perennial plant establishment, on most arid- and semiarid wildlands, except in years with unusually high precipitation. Annual legumes are often used as nurse crops, but they greatly reduced the establishment of *Artemisia californica* in the southern California coastal sage scrub (Marquez & Allen, 1996). Nurse crops are less frequently used on water-limited wildlands, or where soil fertility is limited.

Oats (*Avena fatua*) and barley (*Hordeum vulgare*) are common nurse crops for establishing perennial plants. Common rye (*Secale cereale*) is too competitive for a good nurse crop, and wheat (*Triticum aestivum*) somewhat intermediate. Competition from these nurse crops can be reduced with strategies that: reduce the seeding rate of oats or barley to between 7 and 11 kg ha^{-1}; drill nurse crops and perennial species at 90° angles or in alternate rows; and harvest the nurse crop early (Vallentine, 1989).

On degraded acidic soils, switchgrass improved fertility by adding organic matter, raising pH, elevating cation exchange capacity, and concentrating major nutrients (N, P, K) (Choi & Wali, 1995). It also facilitated the natural recruitment of *Populus* spp. (aspens), *Salix* spp. (willows), and *Betula* spp. (birches) by physically capturing wind-blown seed and acting as a nurse crop for woody plant seedlings (Choi & Wali, 1995).

Planting annual, residue-producing crops during the growing season prior to seeding a perennial species and then directly seeding into the residue is the preparatory crop method. Preparatory crops are effective because they (1) reduce wind and water erosion; (2) reduce evaporation; (3) reduce weed problems; (4) protect young seedlings from sand damage; (4) lessen seedbed temperature extremes; (7) trap snow during winter to increase soil water; and (8) have income potential (from sale of grain) that partially offsets expenses. The preparatory crop method is the most successful seedbed preparation method in the southern US Great Plains. It is most effective under dryland conditions where wind and water erosion are serious hazards. Preparatory crops reduce soil surface drying and crusting following rain. In northcentral Texas, seeding in the dead litter of preparatory crops was 88% successful, while seeding into clean, tilled seedbeds was 67% successful (GPAC, 1966).

In Wyoming, small grain crops are seeded in the spring and perennial grass mixtures are seeded into the stubble the following autumn. This strategy was superior to seeding into hay residues or crimped straw because less mulch was lost to wind or water erosion, treatment costs were 75 to 95% lower, and fewer weed problems occurred (Schuman *et al.*, 1980). The stubble treatment also increased water infiltration more than crimped straw treatments. Sorghums (*Sorghum bicolor*), foxtail or Italian millet (*Setaria italica*), browntop millet (*Panicum ramosum*), Japanese millet (*Echinochloa crusgalli* var. *frumentacea*), and pearl millet (*Pennisetum americanum*) are used as preparatory crops.

It is important that preparatory crops be seeded in rows spaced no more than 50 cm apart and that they not be allowed to set seed, since volunteer plants strongly compete with seeded grasses during the next growing season (Vallentine, 1989). Winter wheat can be used as a preparatory crop by seeding in late spring to prevent vernalization and seed production. Forage sorghums can be planted in late summer, so they will not have time to produce seed before frost.

Herbaceous legumes are used as preparatory crops to (1) provide cover that protects the soil from wind and water erosion; (2) improve soil tilth; (3) reduce weed growth; (4) increase beneficial insects; (5) add nitrogen to the soil; and (6) improve wildlife cover and food. Organic nitrogen contributions to the soil may improve the subsequent establishment of perennial species. Recent crop research findings have suggested the value of herbaceous legumes as a living mulch or preparatory crop in wildlands. When grown with corn (*Zea mays*), alfalfa (*Medicago sativa*) suppresses weeds, supplies all the nitrogen needs, and increases beneficial insects for biological pest control (Grossman, 1990). Subterranean clover (*Trifolium subterranean*) has shown promise as a living mulch for both soybeans (*Glycine max*) and broccoli (*Brassica oleracea*) (Grossman,

1990). Subclovers are useful because they naturally die off in late spring. This provides mulch that reduces weed growth during the summer. They might be grown as a preparatory crop in year one and then warm-season, perennial species could be planted (with or without herbicide-treated strips) into the subclover bed. However, the feasibility of this approach is limited to situations where soil water reserves can reliably support both the subclover and the warm-season plants in the same year.

Strip cropping is a variation of preparatory cropping that has been used in the semiarid portions of the North American Great Plains where wind erosion is a serious hazard. During strip cropping, mechanically fallowed strips (each 10 m wide) are alternately seeded to perennial grasses (Bement *et al.*, 1965). Grass strips alternate with similar sized strips planted to annual crops, such as cotton (*Gossypium hirsutum*), wheat or grain sorghum. Then after the grass strips are established, the previously cropped strips are fallowed one year and planted to grasses the next year. This strategy reduces wind erosion hazards during the entire establishment process compared with planting perennial grasses on the entire area during the same year.

Nonnative tree plantations have been used to alter microenvironmental conditions. These changes allow interior forest species to establish naturally or be planted (Uhl, 1988; Lugo, 1992; Brown & Lugo, 1994; Guariguata *et al.*, 1995). In Puerto Rico, tree plantations improved soil and microenvironmental conditions enough to facilitate the natural immigration of native species (Lugo, 1992). The plantation also accelerated the return of native species by attracting animals that imported seed. Tree plantations in the moist and wet tropics do not remain monocultures (Lugo, 1992), because native trees invade the understory and penetrate the canopy of the exotic species. If site damage is not extreme, native forests replace the nonnative plantation. Where the damage was more extreme, the resulting community is a combination of native species and plantation trees (Lugo, 1992). Nurse crops of fast-growing leguminous trees are believed to be the best strategy for restoring native, dry-forest trees to their former habitat in the US Virgin Islands (Ray & Brown, 1995). Tree plantations also facilitate tropical forest recovery in degraded lands in Costa Rica because they create a favorable microenvironment and reduce grasses that outcompete native trees (Guariguata *et al.*, 1995). Similarly, tree canopies were found necessary for the establishment of native trees in the jarrah forests of Western Australia (McChasney, Koch & Bell, 1995).

Using tree plantations to restore tree richness is effective when managers match species to particular site conditions and overcome limiting factors that prevent the regeneration of native species (Lugo, 1997). After a forest canopy is returned, microenvironmental conditions change and animals that bring seed are attracted. However, some plantation species inhibit native species (Murcia, 1997) and animal transport of seed is not always fully effective. Although tree plantations can effectively recreate favorable environments for native tree species, animals should not be expected to return all species to the area (Parrotta, Knowles & Wunderle, 1997).

Special seedbed considerations

Water-limiting environments, salinized soils, active sand dunes, and mulch applications are unique circumstances that require alternative technologies.

Water-limiting environments

Seedbed management strategies that address water limitations provide significant benefits (Weber, 1986). Water limitations are most common in arid- and semiarid ecosystems, but also occur on severely degraded sites in humid environments and saline soils. Since precipitation in these areas often falls in widely spaced, intense events, most of the water is lost from the site. Water can limit establishment even in relatively high precipitation regions, particularly if soil conditions are poor.

Specialized surface soil modifications that collect runoff water require additional investments, but are the most reliable establishment technique in many areas (Weber, 1986). Some aridland farming systems harvest water from areas treated with latex, asphalt, or wax to improve runoff efficiency (Ffolliott *et al.*, 1994), but those approaches are uncommon on wildlands. The most common strategies for wildlands include some method to harvest or concentrate runoff water or to trap wind-blown snow.

Figure 6.6. Depressions created in an Idaho seedbed by land imprinter. This relatively sheltered microenvironment reduces water loss and can improve seedling establishment in some situations. Photograph courtesy of Warren Clary.

WATER HARVESTING

The direct benefits of water harvesting strategies such as pitting and contour furrowing are generally short lived (Vallentine, 1989). These soil modifications have a finite life that is determined by erosion rate, depth, and precipitation events. However, even with a short lifespan, they can establish long-lived plants that have a lasting, self-perpetuating impact on the site. For example, the positive impacts of favorable microsites were still detectable 20 years after planting on minespoils in India (Jha & Singh, 1992). Water-harvesting techniques that establish shrubs to change microenvironmental conditions and harvest wind-blown soil, nutrients, and propagules may have long-term benefits in arid and semiarid ecosystems (Whisenant, 1995; Whisenant *et al.*, 1995; Whisenant & Tongway, 1995).

Creating depressions in the soil surface to concentrate water increases seedling survival (Figure 6.6) and dramatically increases agricultural productivity in arid ecosystems (Reij, Mulder & Bergermann, 1988). Microcatchments harvest water from within

100 m of collection basins (Boers & Ben-Asher, 1982) and are effective where there is no defined stream channel (Matlock & Dutt, 1986). They are most appropriate in arid regions with high runoff coefficients, with the basin-to-catchment ratio (ratio of water holding area to water harvest area) being determined by slope, rainfall characteristics, runoff rate, and the requirements of planted species. In the northern Negev Desert (99 mm mean annual precipitation) 95% of the *Atriplex halimus* seedlings planted in 32-m^2 microcatchments established, while those only receiving direct precipitation suffered 100% mortality (Shanan *et al.*, 1970). *Atriplex* seedlings established within the basins, but nowhere else. This microcatchment system greatly increased productivity (Shanan *et al.*, 1970). In southern Arizona (150–200 mm mean annual precipitation), microcatchments increased *Cenchrus ciliaris* (buffelgrass) productivity five-fold over a four-year period (Slayback & Cable, 1970). In an arid region near Jodhpur, India, shrub seedlings established more readily when planted in 60 × 60× 60 cm pits with crescent-shaped dikes on the downslope side (Tembe, 1993).

Water harvesting methods do not guarantee success. Seedings in water harvesting environments may still fail during very dry years. Water harvesting is often unnecessary during wet years. However, water harvesting increases seedling establishment and plant production during the years that are neither too dry nor too wet. Like other risk-reduction strategies, water harvesting increases the probability of success, it does not eliminate failure. Deciding which water harvesting strategy (if any) is most appropriate for a particular application requires an understanding of local precipitation patterns and seedling establishment requirements.

SNOWFENCES

Snowfences increase water availability for developing plants by capturing and holding snow and holding it until it melts. Seedling survival increased from 70% to 90% for *Juniperus virginiana* and from 0% to 90% for *Pinus sylvestris* planted within 7.6 m of 1.2 m tall snowfences (60% open) (Dickerson, Woodruff & Banbury, 1976). *Juniperus* seedlings planted near snowfences grew 33% taller than other seedlings. Shelterbelts and shrubs provide similar snow-trapping benefits without the maintenance requirements of fences.

Salinized soils

Salinized soils are the result of altered hydrologic processes in arid and semiarid regions and occur through natural or human-caused (secondary salinization) processes. Where possible, site preparation should repair the altered hydrology causing the problem. That may involve lowering the water table with increased transpiration, deep drainage, deep furrowing, or ridging (Ffolliott *et al.*, 1994).

Site preparation techniques in salinized soils should reduce salt accumulations on the surface and encourage downward movement (FAO, 1989). Two approaches effectively overcame seedling establishment problems on salinized sites in Western Australia (Malcolm, 1991). The first was to grow seedlings in another field or greenhouse and transplant well-developed seedlings. This avoided the serious limitations of seed germination and seedling establishment. The second approach involved planting into naturally occurring microsites where leaching occured, or artificially creating planting niches for seedlings. Commercial equipment can create specialized seedbeds and deposit organic mulches around the seed (Figure 6.7). Direct seeding into specially created M-shaped mounds with organic materials placed in the 'V' portion of the 'M' is effective in western Australia (Malcolm, 1991). This seedbed configuration increased germination and establishment by concentrating rainwater near the seed and leaching the salts. The organic materials reduce crusting, increase infiltration, reduce evaporation, and protect the seed and developing seedling. The best height of the 'M' depends on the circumstances, ranging from saturation to precipitation regimes as low as 175 mm (Malcolm, 1991).

Other effective seedbed preparation strategies improve the soil water, salinity, and temperature conditions. Planting into furrows increased the availability of soil water to developing seedlings and reduced evaporative water losses (Evans *et al.*, 1970). Furrowing also increased the effectiveness of natural precipitation in leaching salts from surface soils. In semiarid areas, seedling establishment was higher in deep furrows than in shallow furrows (Roundy, 1987). Deep plowing improved growth on soils by increasing root penetration, infiltration rate, and leaching depth (Smith & Stoneman, 1970; Sandoval & Reichman, 1971). However, deep plowing and furrowing causes problems in at least two situations: (1) vesicular-crusted, silt loam soils (Roundy, 1987); and (2) sodic subsoils that mix with surface

Figure 6.7. Kimseed contour seeder planting saltbush (*Atriplex*) seed in mounds that reduce the detrimental effects of saline soils. Photograph courtesy of Australian Revegetation Cooperation.

soils during plowing (Mueller, Bowman & McGinnies, 1985). Plowing vesicular-crusted soils can cause sloughing and flowing that leads to excessive seed burial depths (Wood *et al.*, 1982). These sites are effectively planted with land imprinting equipment that press furrows and seed into the soils without plowing (Roundy, 1987). Deep plowing that mixes a sodic horizon with the surface soil can also reduce seedling establishment and growth.

Leaching with low sodium water or leaching gypsum ($CaSO_4$) through the soil improves most sodic soils. The Ca^{2+} from gypsum replaces Na^+ and flocculates clays into aggregates and the acidic anion reduces the pH (Loomis & Connor, 1992). Leach the Na^+ downward with large amounts of water. Since irrigation water is seldom available, using salinity-tolerant species is more practical. Planting salinity tolerant species (halophytes) on salt-affected land is a practical alternative in many situations. Woody plants are commonly used and several saltbush (*Atriplex*) species dominate that usage in arid and semiarid climates. Despite their salinity tolerance, halophytes may require special planting procedures for successful establishment (Malcolm, 1991). Halophytes establish in favorable niches with reduced salinity. These favorable niches occur in specific microsites and certain times. Precipitation temporarily leaches salts out of these sites, creating a short-lived window of opportunity (Malcolm, 1991). This window of opportunity is lengthened with additional precipitation or mulch that reduces evaporation from the soil. Rapidly developing seedlings have significant advantages in this short-lived, favorable environment.

Active sand dunes

Actively moving sand dunes are a hostile environment for developing seedlings, because of sandblasting, uprooting, or burial. In arid and semiarid environments, plants only grow on certain parts of a particular dune type. Some dune types have no vegetation potential in arid environments. For example, transverse, barchan, and seif dunes are never vegetated in arid regions (Floret et al., 1990; Tsoar, 1990; Thomas, 1992). Only the more stable kinds of dunes (such as vegetated linear dunes) are vegetated in arid and humid regions. Thus, it is important to understand the inherent potential for stabilizing specific dune types in a particular climate prior to developing repair options for those sites.

There are many effective techniques for stabilizing active dunes, but the cost of those techniques often exceeds the market value of the land. A practical strategy, for some situations, is to remove livestock and fertilize the existing vegetation until it stabilizes the dunes (Eck et al., 1968). Repeated fertilization may produce the plant biomass necessary to stabilize dune movement.

Even where the soil and climate have the potential for a vegetative cover, simply planting seed or seedlings is rarely effective without some type of protection during seedling establishment. Preplanting barriers provide that protection by reducing uprooting, abrasion injury, and burial (Kavia & Harsh, 1993; Ffolliott et al., 1994). Although complete arrest of sand flow is not possible (Watson, 1990), active dunes may be initially stabilized with petroleum products spread on the dune sand, chemical mulches and sealants, or diversion fences made of wood panels, stone, or soil (Ffolliott et al., 1994). Spraying 1403 or 2807 L oil ha^{-1} on sandy repair sites significantly reduced erosion from coastal dunes in Massachusetts (Zak & Wagner, 1967). Two and one half years after planting, the mulch was still in good condition, no erosion had occurred, and the woody plants were well established. Unfortunately, asphalt mulches usually prevent the establishment of herbaceous species (Zak & Wagner, 1967; Eck *et al.*, 1968). Another study, conducted in Texas, found asphalt mulch patches completely unsatisfactory (Eck *et al.*, 1968). No vegetation established adjacent to the asphalt patches and wind erosion undermined and destroyed the patches with one year.

For practical reasons, it is advisable to construct preplant barriers one year prior to planting. Locally available materials such as twigs and branches, straw, old railroad ties, poles, and earthen ridges create aboveground obstructions perpendicular to the prevailing wind direction. Preplant barriers can be arranged in checkerboard patterns if strong winds often come from two or more directions. Wind velocity, slope, and type of dune all influence the spatial arrangement of preplanting barriers (Ffolliott *et al.*, 1994). Preplant barriers, of partially buried shrub stems, stabilized active dunes near Jodhpur, India (Kavia & Harsh, 1993). Parallel rows were created 2 to 5 m apart with the narrow spacings on the tops of dunes. The more distant row spacings were on the windward sides and interdunal depressions. These preplant barriers significantly increased shrub and tree establishment.

Lasting stabilization of active dunes requires a protective vegetative cover (Brooks *et al.*, 1991). Although herbaceous cover is useful, shrubs and trees are most effective because they capture more wind-blown particles, reduce wind speed at the soil surface, and usually live longer than herbaceous materials. Plants chosen for dune stabilization should have (1) well-developed root systems; (2) tolerance of high wind; (3) the ability to withstand abrasion from moving sand particles; (4) tolerance of rapid soil loss or soil accumulation; and (5) aggressive vegetative reproduction.

Mulches

Seedbed mulches reduce soil erosion (Siddoway & Ford, 1971), lessen temperature extremes, conserve soil moisture, increase seed germination, and increase seedling growth (Zak & Wagner, 1967; Eck *et al.*, 1968; Singh & Prasad, 1993). Because mulches change the nature of the seedbed, the type and amount of mulch affect the species that establish (Luken, 1990; Munshower, 1994). Mulches that improve seedbed conditions include straw, hay, wood chips, shredded bark, peat moss, corncobs, sewage sludge, sugarcane trash, manure, plastic, and synthetic petroleum products (Luken, 1990; Singh & Prasad, 1993). The benefits of mulches appear greatest in arid environments (Winkel, Roundy & Cox, 1991; Singh & Prasad, 1993; Roundy, Abbott & Livingston, 1997) and where weed competition is a serious obstacle. In New Jersey, dense oak (*Quercus* spp.) litter increased the size of woody plant seedlings by reducing herbaceous plant growth (Facelli & Pickett, 1991). Seedbed mulches increased the growth of Douglas fir (McDonald, Fiddler & Harrison, 1994) and one-seed juniper (*Juniperus monosperma*) (Fisher, Fancher & Aldon, 1990) by reducing competition from other species. However, where weed problems are less severe (e.g., mined soils) and contour furrows will control erosion, mulches may not be necessary (McGinnis, 1987).

Organic materials provide nutrient supplements and improve the soil's resource retention capacity (Table 3.3). To produce mulches in low moisture situations, leaves of rapidly decomposing species should be mixed with more slowly decomposing leaves. This mixed mulch improves water conservation and slows N release by the mulch, to increase the period of available N to planted species (Seneviratne, Holm & Kulasooriya, 1998). Tacking agents bind organic materials (e.g., wood fibers) to the soil. Chemical tacking agents require caution since they can inhibit seed germination (Sheldon & Bradshaw, 1977). Mechanically crimped mulch is partially inserted into the soil and does not blow or wash off the soil surface.

Gravel, stones, rocks, and even oil are useful for certain applications. Gravel mulches increase germination under water-limiting conditions, unless they are too deep (Winkel *et al.*, 1991; Winkel *et al.*, 1993). In arid regions of India, gravel mulches reduced moisture losses from planting sites and were more stable during high winds (Mertia, 1993). Gravel, stone, and rock mulches increased seedling establishment during natural or artificial recovery of disturbed sites in arctic Alaska (Bishop & Chapin, 1989b; Bishop & Chapin, 1989a). Where readily available, rock mulches shield woody plant seedlings from temperature extremes and provide some protection from herbivory. Three or more 10–20 cm diameter rocks arranged around seedlings have enough thermal mass to provide thermal buffering and reduce evaporation (Bainbridge, Fidelibus & MacAller, 1995).

7

Planting

Introduction

Planting seed, plant parts, or entire plants are common techniques for introducing species into repair sites. Direct seeding puts seed into the soil with some control over depth, density, and spacing. Broadcasting seeding scatters seed over the soil surface. Planting entire plants (wildings, containerized seedlings, or bare-root seedlings) or plant parts (stolons, rhizomes, bulbs, corms, or stem sections) is more reliable in harsh environments, but has greater expense. Effective planting strategies contain information on the most effective planting time, planting rate, planting depth, and the most appropriate equipment. The goal is to increase establishment success by placing the plants or seed in the soil at the optimum time and under the ideal conditions.

Direct seeding

Direct seeding has several advantages compared with transplanting. Most repair programs prefer seed to seedlings because seed are less expensive, easily stored and transported, and more readily collected or purchased. This flexibility, and the relatively low cost, make direct seeding the preferred planting method for many situations. It is particularly useful where access, terrain, or soil conditions make transplanting whole plants difficult, too expensive, or impossible (Barnett & Baker, 1991). When direct seeding, the most important considerations involve seed preparation, planting time, planting depth, seeding rate, and selecting the most appropriate seeding method.

Seed preparation

Many species require some type of seed treatment before planting. These treatments break natural dormancy mechanisms that delay germination, or inoculate the seed with bacteria to encourage symbiotic, nitrogen-fixing relationships.

BREAKING SEED DORMANCY

Even with ideal seedbed conditions, some seed germinate slowly or not at all after planting. Under natural conditions, seed dormancy prevents the seed from germinating at times when the developing seedlings have less chance of surviving. Two categories of dormancy mechanisms cause germination delays. The first (internal dormancy), is caused by the incomplete digestion of fats, proteins, and other complex insoluble substances stored in the seed (Smith, 1986). These compounds convert to simpler, organic substances (i.e., sugars and amino acids) that translocate to the embryo. Under natural conditions, for some temperate trees, these conversions occur during cool, moist conditions. We simulate those conditions by placing the seed in moist, peat moss or sand stored just above freezing (Smith, 1986). This is cold-moist stratification. Other species require warm-moist or cold-dry stratification (Steffen, 1997).

Impervious seed coats that exclude oxygen or water from the embryo cause the second type of seed dormancy (Smith, 1986). Under natural conditions, alternating extremes of temperature, microbial action, and abrasion against soil particles erode the seed coat and increase permeability. This is scarification, or the process of breaking the impervious seed coat to allow the uptake of water and oxygen. Mechanical abrasion or chemical softening of the seed coat will remove this dormancy mechanism. Tumbling seed in a container with coarse, sharp sand, or some other abrasive material, mechanically scarifies seed with hard seed coats. Chemical scarification involves soaking the seed in hot water, sulfuric acid, or nitric acid to break the hard seed coat. Herbivores scarify seed when they pass through the digestive system.

INOCULATION

Inoculation attaches appropriate bacteria to seed (usually legumes) to assure nodulation and symbiotic N fixation. It is usually done, prior to planting, by coating the seed with a water-based slurry of bacteria, and a peat carrier that is enriched with sugars, gums, and complex polysaccharides (Heichel, 1985). This mixture provides nutrition for the bacteria and aids in sticking them to the seed. Although commercial stickers are readily available, soft drinks are also used as the sticker (Steffen, 1997).

Inoculation is not necessary if the same legume is, or has recently been, abundant on the site. Those soils should contain nodule bacterial populations to adequately nodulate seeded plants. However, since proper inoculation adds relatively little to the cost, it seems prudent to inoculate legumes before seeding. Native nodule bacteria are less persistent in soils that are strongly acidic or alkaline, prone to nutrient deficiencies, subject to high temperatures or moisture stress, exposed to pesticides or fertilizers, or have high N levels (Heichel, 1985). Thus, inoculation is important under those conditions. Inoculum has a limited shelf life, as indicated by the expiration date on the label.

Planting time

Poorly timed seedings are less likely to succeed. Since germination and seedling establishment require ample soil moisture and favorable temperatures, the best time for direct seeding is just prior to the longest period of favorable growing conditions. Unfortunately, unanticipated events and extreme environmental conditions cause seeding failures, even when seeding occurs at the best time (Ries & Hofmann, 1996). No planting time guarantees success; we pick planting times that most likely precede good growing conditions. In most areas, we identify that time by considering seasonal temperature and precipitation patterns.

Climatic patterns create predictable differences in the best seeding time. Cool-season species are most effectively seeded in Mediterranean climates (hot, dry summers and cool, wet winters) in late autumn prior to the first rains (Heady, 1975; Lefroy, Hobbs & Atkins, 1991). They germinate in late autumn and the seedlings have time to grow before winter temperatures arrive. In temperate, continental climates (e.g., Mandan, North Dakota) cool-season species established more reliably when seeded in the fall, whereas warm-season species established best with late winter to mid-spring plantings (Ries & Hofmann, 1996).

Autumn seeding is most suitable where most of the precipitation occurs as winter snow and the primary growing season occurs during a brief, early spring period. The aim should be to plant late enough so the seed do not germinate before they are covered with snow. The seed then germinate immediately after snow melt and take full advantage of this short growing season before the onset of hot, dry summers. This is critical because it is the best growing period of the year and every day is important. Spring seeding is less effective where most of the precipitation falls as snow. By the time equipment can get into the field, the most important part of the growing season is over. Thus, by the time snowmelt allows planting, the best growth period is over and soil water reserves are much lower.

There are several disadvantages to autumn seeding: (1) there is more time for birds and rodents to remove seed; (2) seed that imbibe water and then dry out will probably die; and (3) there is more potential for weed problems in the spring (Lefroy *et al.*, 1991). Other circumstances suggest alternative seeding times. Wildlife migration patterns may prompt changes to avoid destruction of newly planted sites. Autumn planting dates favor dormant seed that require cold stratification.

In England, many deciduous forest species were most effectively planted in the spring, but some species require autumn planting to meet cold requirements (Harmer & Kerr, 1995). Planting in tropical, monsoonal climates without distinct warm or cold seasons should occur just prior to rainfall and the longest period of good growing conditions. Both spring and autumn planting may be possible where temperature and precipitation patterns are suitable (e.g., humid, temperate to subtropical areas). Although spring planting is more common in these areas, autumn plantings may reduce interference from weeds. Sowing seed at different times may insure that at least some of the seed germinate even under unusual conditions. Multiple germination periods may insure a wider range of genetic variation is represented in the established population (Packham *et al.*, 1995).

While seeding into a weed-free seedbed is most effective, it is not always possible. Seeding into competitive environments requires the use of competitive species and a seeding time that is most advantageous to the seeded species. The relative time of introduction into a community dramatically affects the competitive outcome. For example, planting

Lolium perenne and *Plantago lanceolata* at the same time produced a crop in which *Lolium* made up 80% of the total dry weight (Harper, 1977). Planting *Lolium* three weeks earlier than *Plantago* increased the *Lolium* contribution to 90%, but planting *Plantago* three weeks earlier resulted in a mixture in which *Lolium* comprised only 6% of the final dry weight. Neighbors immediately dominate new seedlings. Thus, early-emerging seedlings have an advantage and continue to increase their ability to capture resources at the expense of late-emerging seedlings (Ross & Harper, 1972).

Seasonal accessibility of the repair site to planting equipment affects the choice of seeding dates. Accessibility is controlled by precipitation patterns and soil texture. Spring seeding is most effective with warm-season species in climates where the primary growing season occurs in late spring and/or summer. Sandy, well-drained soils are prepared and planted in early spring. This not only gets the seed in the ground at the proper time, but spring cultivation reduces competition from many weeds. However, heavy, clay soils that cannot be reliably worked, when wet, in the spring may need to be seeded in the fall (Heady, 1975).

Seeding rate

Seeding rate is the number of planted seed per area (kg ha^{-1}). The seeding rate for mixtures is either stated as a rate for each species or for the entire mixture. Seeding rates should be high enough to produce the desired vegetative density and cover, but low enough to be cost effective and limit self-thinning. Seeding rates are adjusted to many variables, including number of seeds per kilogram, purity, germination percentage, seedbed conditions, growth habits, management objectives, and seed costs. It is necessary to know whether recommendations are based on Pure Live Seed (PLS) or bulk seed (see Equations 5.1 and 5.2).

It is advisable to consider local knowledge of species adaptations, mixtures, and seeding rates. In the absence of that information, seeding rate guidelines are useful. When planting grasses, one of the more widely used guidelines is to plant 200–400 PLS m^{-2} or enough to establish 10 plants m^{-2}. Herbaceous seeding rates often vary with seed size. Vallentine (1989) recommended drilling approximately 100 PLS m^{-2} for large seed (less than 143 000 seed kg^{-1}); 200 PLS m^{-2} for medium sized seed (143 000 to 1.1 million seed kg^{-1}); 300 PLS m^{-2}

for small seed (1.1 to 2.2 million seed kg^{-1}); and 400 PLS m^{-2} for very small seed (over 2.2 million seed kg^{-1}). These resemble other recommendations of 5.0–10.0 kg ha^{-1} for large seeded plants (Mueggler & Blaisdell, 1955; Launchbaugh, 1970; Hull, 1972), or 30 g to 3 kg ha^{-1} for small seeded species (*Eragrostis, Panicum, Poa*, and most herbaceous legumes) (Kilcher & Heinrichs, 1968). Prairie restoration guidelines call for higher seeding rates, ranging from 400 to 600 PLS m^{-2} (Diboll, 1997).

Although seeding 100–400 PLS m^{-2} may seem excessive, losses during germination, emergence, and establishment require these rates. Only 10–30% of planted seed produce seedlings, and less than 50% of those survive (Decker & Taylor, 1985; Vallentine, 1989). Although these establishment rates (5–15%) seem low, complete failures also occur. These guidelines provide useful first approximations, but they must be adapted to local conditions. For example, it is unrealistic to expect 10 established plants m^{-2} in arid ecosystems. Several situations may require changes in these guidelines. General seeding guidelines may require modification after considering (1) site, climatic or seedbed conditions; (2) seeding method; (3) potential for weed problems; and (4) erosion potential.

It is advisable to increase seeding rates to overcome high mortality rates, rapidly stabilize soil resources, or to reduce weed problems. Harsh sites and poorly prepared seedbeds require a higher seeding rate because we expect low success rates. For similar reasons, we double seeding rates when broadcast seeding. Broadcast seeding requires higher seeding rates to compensate for uneven seeding depth, uncovered seed, and seed loss to rodents and birds (Vallentine, 1989). Sparse stands of small seedlings do not stabilize soils, so higher seeding rates accelerate site stabilization. Higher seeding rates reduce the detrimental effects of weeds (Vallentine, 1989; Stevenson, Bullock & Ward, 1995). Weed control increases at higher seeding rates and with narrow row spacings. The effect of narrow row spacings improves when combined with measures to improve the competitive ability of the seeded species, such as a single weed control operation. The number of seed drilled into good sites may be lower than 200 PLS m^{-2}, but should not be reduced below the number needed to exclude weeds from the site (Munshower, 1994). Although seeding rates are site specific and may differ for each circumstance, local knowledge may provide good estimates. Without more site specific information, 200 PLS m^{-2} is a useful first approximation when drilling seed into well-prepared seedbeds.

Predictably, high establishment allows the use of lower seeding rates. Favorable seedbeds with deep well-prepared soils, gentle slopes, and a good moisture regime tolerate lower seeding rate. Satisfactory establishment of conifer seedlings required 3.3 kg ha^{-1} of high quality seed on undisked sites, but 1.5 kg ha^{-1} was sufficient on well-prepared seedbeds (Stoddard & Stoddard, 1987). In Western Australia, seeding rates of 0.4–1.0 kg ha^{-1} were needed to reestablish native woody species on previously cultivated lands (Lefroy *et al.*, 1991). These rates equate to seeding 200 g to 500 g km^{-1} of row with rows spaced 5 m apart (2 km of row occupies 1 ha). Lower seeding rates are appropriate where the objective is to supplement existing populations. Species with aggressive vegetative expansion (e.g., strong stoloniferous or rhizomatous habits) perform well with reduced seeding rates.

SEEDING RATES AND COMPETITIVE INTERFERENCE

Seeding rates influence the competition between seeded plants and weeds. Consequently, seeding rates can be used to increase the survival and growth of seeded plants. Plant yield per unit area increases with density until each increase produces progressively lower yields. Eventually, higher seeding rates produce no yield increase because available resources limit the entire population. This upper limit (or carrying capacity) of the density–yield relationship is described by the law of constant yield (Kira, Ogawa & Shinozaki, 1953). As an example, low densities of perennial ryegrass (*Lolium perenne*) produced robust plants with many tillers, but the number of tillers per plant decreased rapidly as the seeding rate increased (Weiner, 1990). With higher seeding rates, seedling size becomes progressively more variable, with smaller plants being more susceptible to mortality. Thus, although yield per unit area increased at higher seeding rates, fewer seedlings survived.

Since growth, reproduction, and mortality continue after the initial seedling establishment phase, the yield, and density of seeded plants change over time. In weed-free, well-prepared seedbeds, there are no long-term yield or density differences over a wide range of seeding rates. For the first three years, crested wheatgrass seeded at 4.4–52.8 kg ha^{-1} had higher plant densities on plots with higher seeding rates (Mueggler & Blaisdell, 1955). By the sixth year after planting, plant densities were similar across all seeding rates. In England, higher seeding rates rapidly eliminated weeds on chalk grasslands, but even quite low seeding rates were successful in establishing seeded species and eventually eliminating weeds (Stevenson *et al.*, 1995). Two seasons after planting, all seeded plots resembled native chalk grasslands, while unseeded control plots were dominated by weeds and showed signs of developing into species-poor grasslands (Stevenson *et al.*, 1995). Unfortunately, this pattern of low seeding rates eventually producing as much vegetative cover as high seeding rates is not universal. In Kansas, native grass seedings that began slowly were never able to develop fully because of weed competition (Launchbaugh, 1970). Consequently, it seems prudent to restrict low seeding rates to seedbed environments with few competing plants.

PLANTING VERY LOW SEEDING RATES

Rare and expensive seed are usually planted at very low seeding rates. Widely scattered, low-density plantings are used in the hope that the species will subsequently increase its population density. Rare and expensive seed require precision seeding methods to accurately apply small amounts of seed. Small, hard seed settle to the bottom of seed boxes, resulting in poor distribution across the planting site. Direct-seeding equipment seldom plants small amounts of seed efficiently. Bulking agents help distribute small seed or small amounts of seed over large areas. Commonly used bulking agents include sawdust, rice hulls, branflakes, wheatbran, cornmeal, vermiculite, or some other free-flowing inert material. These materials may become sticky when damp and restrict the flow of seeds through planting equipment. Sand is effective, but has abrasive properties that damages equipment. These bulking agents improve the flow of seed through seeding equipment and the distribution of seed across the site.

Seeding depth

Optimal seeding depth is a compromise between enabling a seed to emerge from the soil and trying to provide a reliable environment for germination and growth. One general rule that incorporates this tradeoff between seed size (e.g., energy reserves) and optimum seeding depth suggests that seed should not be planted deeper than seven times the seed diameter, with four to seven times the seed diameter being the preferred depth (Welch *et al.*, 1993). This optimum seeding depth varies with the period of available water in the seedbed, seed size, germination requirements, and seedling vigor (Roundy & Call, 1988).

The soil surface is easily wetted from light precipitation events, but dries out just as rapidly. Seed germinating from light rains on soil, with little stored water, are very susceptible to desiccation, and whereas more deeply buried seed are buffered against rapid water loss to the atmosphere. For example, the seedling establishment of Indian ricegrass (*Oryzopsis hymenoides*) which grows in arid environments, increases with depth up to 5–8 cm in uncrusted, sandy soils (Vallentine, 1989; Young *et al.*, 1994). However, deeply buried seed require more energy to extend their photosynthetic area above the soil surface. Only seed with large energy reserves are able to emerge and establish when deeply buried (Harper, 1977). Because very small seed are less likely to emerge if buried below the surface, they often require a light stimulus for germination. These small seed should be sown on the surface of a firm seedbed and kept moist until after germination occurs (Steffen, 1997). Lightly pressing them onto the soil surface will insure good seed-to-soil contact. This light requirement is absent from most large seed (Harper, 1977) and they are more capable of emerging when deeply buried. The advantages of deep burial are a more constant moisture and temperature environment.

Small seed, such as *Poa, Eragrostis*, and *Sporobolus* should be planted 3–7 mm deep. Very small seed should be broadcast onto a well-tilled seedbed without covering the seed. For example, Lehmann lovegrass is most effectively planted on the soil surface, since it does not emerge when planted deeper than 3–4 mm (Cox & Martin, 1984). Species with very small seed are effectively planted by broadcasting the seed over a rough soil surface since they establish more readily when placed on the soil surface (Roundy *et al.*, 1993). They often establish without additional soil preparation, but may be lightly covered with a drag, trampling by animals or a light harrow (Heady, 1975). Small-seeded shrubs (e.g., *Ceratoides lanata*) should be planted at 1–2 mm, since they will not emerge if planted below 12 mm (Springfield, 1970). *Artemisia tridentata* ssp. *vaseyana* also has a very small seed that should be seeded no more than 3–4 mm deep (Jacobson & Welch, 1987).

As with any general rule, there are exceptions and situations where other strategies are more effective. For example, in humid environments, there is less advantage to deep planting and good grass establishment is obtained with a firm seedbed and just enough loose surface soil to cover the seed (Decker & Taylor, 1985). In general, deeply buried seed emerge more readily from sandy soils than clay soils. It is more critical to firm sandy surfaces to obtain optimum seed-to-soil contact.

Drill seeding

Drilling is the best way to obtain uniform seed distribution and seeding depth, providing the seedbed is not excessively soft and fluffy. Drilling often produces uniform stands of seeded plants during the first years after planting. Properly used (Table 7.1), the better rangeland drills can operate on rough, rocky ground, seed at the proper rate, handle many types of seed, and pack soil over planted seeds. Drilling creates small furrows that retain water and provide shelter from sun and wind. This can provide the critical margin of difference in many arid and semiarid landscapes.

Drill seeding is more effective than broadcast seeding for several reasons. Drilled seed are less susceptible to depredation by birds and rodents than broadcast seed (Nelson *et al.*, 1970). Drilled seed remain in a relatively favorable moisture and temperature regime. Broadcast seed are exposed to rapidly fluctuating conditions. These conditions cause frequent starting and stopping of germination and growth. In one study, drill seeding produced good stands of seven different grass species, whereas only one species produced good stands when broadcast seeded (Nelson *et al.*, 1970). Drilling produced three to seven times more grass than broadcast seeding in another study (Hull, 1959). Drilling 6.6 kg PLS ha^{-1} was superior to broadcast seeding 12 kg PLS ha^{-1}.

Grain drills designed for cultivated farmland are poorly adapted to common wildland situations (Young & McKenzie, 1982). They are effective on well-cultivated sites without debris. Uneven seedbeds with rocks and/or woody debris exclude or damage most grain drills. Well-designed rangeland drills have rugged components and high clearance for rough sites (Figure 7.1). Many rangeland drills operate effectively in moderate amounts of surface litter, and most operate effectively in burned areas, open stands of less competitive annuals, and cropland

Table 7.1. *Common mistakes encountered when drill seeding, problems created, and solutions*

Mistake	Problem	Solution
Drill not designed for wildland terrains	Farm-type seed drills that are not designed to operate over rough terrain, rocks, and woody debris require frequent repair	Use equipment designed for rough terrain or restrict use to well-plowed fields without rocks or wood obstacles
Drill not designed for fluffy seed	Fluffy type seed are not pulled into drop tube by picker wheel in seed box, so no seed are planted	Use equipment designed for fluffy seed. Use smooth seed that will flow through more traditional drills
Planting at wrong depth	Seed planted too deeply may not emerge. Seed planted too shallow are more susceptible to desiccation	Use depth bands on disk openers to insure proper depth for soil being planted. Seedbed may need to be firmed
Packing fluffy seed into seed box	Packed seed are not mixed well in the seedbox and may not be pulled into the drop tube or planted	Place seed in seedbox without packing
Not checking drop tubes	Seed (especially fluffy seed) can become clogged up in drop tubes – preventing any further seed from being planted	Check regularly to insure seed are falling into drill furrow. Clean out drop tubes to remove any obstructions (spider webs, bird nests, rodent nests)
Planting too rapidly	Many seed types do not flow smoothly through equipment, so they are not planted evenly	Plant more slowly. Use appropriate equipment for the seed being planted

Figure 7.1. Rangeland drills are able to operate in rough terrain and can plant smooth and fluffy seed simultaneously. They create small furrows for each row, plant seed at controlled depths, and firm soil over the seed.

stubble without additional seedbed preparation. Although designed to operate over slopes, rocky soils, and woody debris, rangeland drills are more effective when used on well-prepared seedbeds.

Rangeland drills usually plant grass into rows 15cm and 35 cm apart. Total forage production, after full establishment, is not greatly affected by row widths between 25 cm and 45 cm (Vallentine, 1989). Suggested row widths are wider on semiarid sites and narrower on mesic sites. Several studies indicate a trend of higher productivity from narrow spacings immediately after establishment to no difference or even higher yields from wider spacings in later years (Sneva & Rittenhouse, 1976; Leyshon, Kilcher & McElgunn, 1981). Narrower rows tend to (1) provide better weed control; (2) produce plants with greater proportions of leaf versus stem tissue; (3) provide greater soil stabilization; and (4) have more resistance to trampling (Vallentine, 1989).

INTERSEEDING

Interseeding involves seeding herbaceous plants into an existing stand of herbaceous vegetation. It is most appropriate where (1) erosion hazards are high; (2) complete seedbed preparation is impractical; or (3) the existing vegetation should be supplemented rather than completely replaced. Erosion hazards are reduced during interseeding, compared with complete seedbed preparation and seeding, because the site remains at least partially protected before, during, or after interseeding. Interseeding spreads seed and costs over a larger area, compared with complete stand replacement. Adding herbaceous legumes to an existing stand of grass improves forage quality. Planting dates are similar for interseeding and regular seeding approaches. Seeding rates of one-third to one-half full seeding rates are typical (Vallentine, 1989). Establishing 0.6–1.2 alfalfa plants (*Medicago sativa*) m^{-2} is recommended when interseeding South Dakota grasslands (Rumbaugh *et al.*, 1965). In the North American west, interseeding is used to establish woody plants on big game winter range (Pendery & Provenza, 1987).

Since established vegetation has a strong advantage over establishing seedlings, it is seldom advisable to drill or broadcast directly into established stands of perennial plants. Even strong competitors establish better without competition. Competition from established vegetation can be reduced by: (1) direct seeding into severely depleted upland sites; (2) planting cool-season species into established warm-season vegetation where sufficient autumn and/or early spring precipitation occurs; and (3) using mechanical or herbicidal methods to kill strips of established vegetation.

Several furrow opener designs are effective when interseeding wildland ecosystems. One design used rotary blades to break up the sod in front of each double disk opener (Smith *et al.*, 1973). Most interseeders use a furrow opener to remove a strip of sod from each row and then plant, cover, and pack seed into the opening. The better interseeders have good control over stripping depth, seeding depth and packing. The most effective width of the controlled strip depends on the vigor of the remaining vegetation, soil moisture, and competitive ability of the interseeded species (Vallentine, 1989). Furrows 20–25 cm wide, 5–8 cm deep, and 100 cm apart are common. Interseeding into highly competitive vegetation or drier environments requires wider cleared strips. Arid sites benefit from increased scalping widths and summer fallowing. The additional scalping width and fallow period increased soil water storage at planting time (Bement *et al.*, 1965). The distance between furrow centers commonly varies from 0.7 m in mesic meadows, to 1 m on subhumid rangelands to 2 m on semiarid rangelands. The depth of the cleared strip influences interseeding success. Shallow scalping may not adequately remove existing vegetation. Deep scalping may plant the seed too deep (particularly on sandy soils).

Rough topography, stoniness, or erosion hazards prevent cultivation on many sites. We can create planting strips with herbicides, on these sites, by attaching spray nozzles in front of each row of the seed drill (Waddington, 1992). Spraying contact herbicides on established vegetation reduces competition for the developing seedlings (Vough & Decker, 1983). Paraquat ($1,1'$-dimethyl-4-4'-bibyridinium ion) controlled established grass when interseeding herbaceous legumes in a maritime climate and enough soil moisture (Bartholomew, Easson & Chestnutt, 1981; Vough & Decker, 1983). Herbicides such as glyphosate (N-[phosphonomethyl]glycine), that translocate and kill the entire plant provide longer control than paraquat, which only kills the sprayed tissue (Waddington & Bowren, 1976). In northeastern Saskatchewan, killing strips of existing vegetation with glyphosate and planting legumes in the sprayed strips was an effective technique for adding legumes to perennial grass pastures (Malik & Waddington, 1990). Interseeding into mechanically cleared strips only involves drilling into the soil, but interseeding sprayed strips requires that the drill penetrate the sprayed material before planting seed. Drills with independently suspended disk openers and enough weight to penetrate the herbaceous material are most effective (Waddington, 1992). They must also have depth bands to control seeding depth and a mechanism to pack the soil around the seed.

Inadequate control of competing vegetation causes most interseeding failures, but other problems occur. Interseeding heavy clay sites is difficult because of soil crusting and wet soil sticking to the equipment. Recently interseeded sites should be protected from grazing animals, because new seedlings are more palatable than existing vegetation. Resident rhizomatous and stoloniferous species will rapidly invade seeded strips.

Broadcast seeding

The major advantages of broadcast seeding are its increased speed and reduced cost. Its major disadvantages are (1) lack of spacing and control over stand density; (2) loss of seed to seed predators; (3) reduction in seed germination and establishment compared with drilling; and

Figure 7.2. Small broadcast seeder mounted on the back of an all-terrain vehicle.

(4) a requirement for higher seeding rates to compensate for reduced germination and increased predation. Seeding into a well-prepared seedbed, covering the seed, and firming the soil to increase seed-to-soil contact reduces these problems.

Hand-operated spreaders are effective on small sites, but types of equipment are more effective for broadcasting seed over larger areas. Broadcast seed (or fertilizer) spreaders mounted on aircraft, farm tractors, small trucks, or all-terrain vehicles (Figure 7.2) easily scatter seed over large areas. Broadcasting seed hay over well-prepared seedbeds provides the seed and a ground cover to increase establishment. Hydroseeding techniques apply seed in a high-pressure water stream over sites that are not accessible to conventional equipment.

SEEDBED REQUIREMENTS

Few broadcast seed germinate and establish unless covered with soil. Although broadcast seeding does not cover the seed, subsequent operations to cover the seed will increase success. Dragging chain, pipe, trees, or other objects over seeded sites will cover many of the seed. Using a cultipacker or other type of roller to cover seed and firm the

Figure 7.3. Kimseed camel pitter creating and seeding pits in a bare, crusted soil. Water harvesting approaches can greatly improve seedling establishment in this situation. Photograph courtesy of Australian Revegetation Cooperation.

soil after broadcasting is effective on freshly plowed seedbeds, if soil compaction is not excessive above the seed (Vallentine, 1989). Heavy concentrations of livestock sometimes increase seedling establishment by trampling and covering the broadcast seed (Howell, 1976), but this approach might not be practical on large areas.

Seed that are broadcast onto recently plowed or ripped soils with 5–8 cm of loose soil are often covered by soil sloughing or soil movement during the first rainfall event (Jordan, 1981; Holechek, Pieper & Herbel, 1989). Commercially available equipment can create small pits on gently sloping, crusted seedbeds (Figure 7.3). The next rainfall event covers seed placed in these pits with loose soil. Small-seeded species are more successful when broadcast onto poorly prepared seedbeds than large-seeded species. Small seed broadcast onto rocky or rough, loose seedbeds fall into suitable safe sites without a covering operation. Rocks provide cracks and other small areas that trap small seed in contact with the soil, allowing germination and growth. Mulches also improve seedbed conditions for broadcast seed when applied after seeding. However, the mulch must not contain seed of competing species.

Inadequate weed control is a major obstacle when broadcast seeding into poorly prepared seedbeds (Nelson et al., 1970; Downs et al., 1993). Broadcast seeding bunchgrasses was less effective in Washington state than drilling because rodents and birds consumed many of the seed (Nelson et al., 1970). Within six weeks, 98% of the wheatgrass seed was lost to rodents. Rodenticides significantly reduced seed predation. Chapter 4 described methods of reducing seed losses to birds and rodents.

AERIAL SEEDING

Aircraft can be used to broadcast seed (1) over very large areas (Barnett & Baker, 1991); (2) where the seed must be applied rapidly (Barro & Conard, 1987); and (3) in remote or inaccessible areas (Prasad, 1993). Properly calibrated helicopters can broadcast tree seed over 600–1200 ha each day (Barnett & Baker, 1991). Aerial seeding is often used in the western US to stabilize slopes denuded by wildfires (Barro & Conard, 1987). As with other types of broadcast seeding, aerial seeding is most effective when seed are applied onto plowed seedbeds (Hull et al., 1963; Nelson et al., 1970; Prasad, 1993), but may also be effective on recently burned sites. Large-scale aerial seedings conducted with compressed earthen pellets containing seed found no advantage compared with using pelleted seed on unprepared seedbeds (Hull et al., 1963).

HAY MULCH SEEDING

Hay mulch seeding involves spreading seed-containing hay over a well-prepared seedbed. It is a favored technique for restoring native species and genotypes because it is the only way to obtain seed of some species. However, since each species produces seed at a different time, many species are absent, or underrepresented, from a single hay harvest. The hay should be cut when the important species are at an optimum stage of maturity, and then raked, dried, and stacked. Drying prevents mold or 'heating' in the stacks or bales so the stacks can be stored (Vallentine, 1989).

Apply seed hay prior to the optimum seeding time for the dominant (or preferred) species within the hay. Spreading the hay by hand is labor intensive and most practical on small sites. Commercial chopper-spreaders are available to shred and apply hay over larger areas. Typically, at least 2000 kg hay ha^{-1} are required; double that rate on highly erosive sites (Vallentine, 1989). The hay may require anchoring

Figure 7.4. Cultipacker-type seeder operating on recently plowed strip in south Texas. This equipment creates many small depressions but requires a well-tilled seedbed.

where wind or water can displace the hay. Commercial hay crimpers, disking, vertically oriented coulter blades, or short-term trampling by livestock are effective at anchoring hay to the soil.

Seed hay supplies seed, improves microenvironmental conditions, conserves water, and reduces soil erosion. Hay mulch seeding has long been used to heal blowouts in sandy areas by stopping sand movement and establishing a permanent vegetative cover (Vallentine, 1989). Sandy soils require little or no seedbed preparation before spreading the hay, if there are few competing plants. However, most situations (especially heavier textured soils) require good seedbed preparation and should be plowed immediately prior to applying the seed hay.

CULTIPACKER-TYPE SEEDERS

Cultipacker-type seeders use seed-metering boxes to drop seed between two heavy, corrugated rollers. The first roller firms the freshly plowed soil surface into small, shallow depressions, and seed drop onto the surface. The second roller firms the seed into new depressions (Figure 7.4) with approximately 0.5–2.5 cm of soil over the seed.

Although cultipacker-type seeders operate on most surfaces, wet clay sticks to the equipment and renders it unusable.

HYDROSEEDING

Hydroseeding (also called hydraulic seeding) broadcasts seed in a high-pressure water carrier. The seed are mixed with water, fertilizer, mulch, and a binder (called a stabilizer or tackifier). Hydroseeding has the potential to rapidly stabilize easily eroded sites and sites that are not accessible to conventional seeding equipment (due to excessive slope or other restrictive soil conditions). It has been recommended for use on high priority mine sites (Munshower, 1994), roadways (Carr & Ballard, 1980), unstable sand slopes (Sheldon & Bradshaw, 1977), and reservoir drawdown zones (Fowler & Maddox, 1974), but is too expensive for most wildland situations.

Despite the advantages of hydroseeding, there are numerous situations where it was less effective than conventional seeding techniques (Sheldon & Bradshaw, 1977; Roberts & Bradshaw, 1985; Munshower, 1994). Hydroseeding problems are often associated with (1) poor seed-to-soil contact when the seed are mixed and applied with a mulch (Roberts & Bradshaw, 1985; Munshower, 1994); (2) fertilizer toxicity (Roberts & Bradshaw, 1985); (3) binder inhibition of germination (Sheldon & Bradshaw, 1977); or (4) reduced infiltration caused by mulches and binders (Sheldon & Bradshaw, 1977; Roberts & Bradshaw, 1985).

When hydroseeding, the seed should not be applied with a hydromulch and/or binder because many of the seeds will be suspended above the soil as the hydromulch and binder dry (Munshower, 1994). Although these seed may germinate, the seminal roots are unable to penetrate the soil and the seedlings become desiccated and die. Hydromulch can be applied over the top of hydroseeded sites with much success (Munshower, 1994).

Some fertilizers are toxic to germinating seedlings, and binders may inhibit germination (Roberts & Bradshaw, 1985). Fertilizer toxicity varies with site and climatic conditions. While some studies found single applications of seed and fertilizer just as effective as sequential applications (Fowler & Maddox, 1974; Carr & Ballard, 1980), others recommended fertilization be delayed until after germination (Roberts & Bradshaw, 1985). Binders reduce mulch losses to wind or water erosion. One comparison found that none of the chemical binders increased seedling establishment and some significantly decreased establishment (Roberts & Bradshaw, 1985). Binders can reduce the infiltration of water into the soil. Mulches are most effective at increasing seedling establishment when long-fiber, flexible materials (Roberts & Bradshaw, 1985) are applied after seeding (Munshower, 1994).

Transplanting

Wildings, container-grown seedlings, bare-root seedlings, cuttings, and sprigs are transplanted with shovels or specially designed equipment. Transplanting wildings, container-grown seedlings, or bare-root seedlings increases establishment success compared with direct seeding. In the most unpredictable and inhospitable environments, 50–75% of all direct seeding operations fail. In the Mohave Desert, where direct seeding operations seldom succeed, transplanting whole plants is the most viable alternative (Holden & Miller, 1993).

Hand tools such as a dibbles, planting bars, shovels, post-hole diggers, or augers are used to plant tree and shrub seedlings (Figure 7.5). Tractors and tree planters are effective, particularly where labor is expensive and large areas must be planted during short planting periods. Tree planters are more effective where there are few obstacles such as trees, gullies, stumps, or rocks. Under good conditions, this equipment plants 1000 seedlings per hour (Stoddard & Stoddard, 1987). While rapid, economical, and effective, these straight-line planting patterns do not recreate natural-looking communities. In the southeastern US, planting seedlings by hand skips fewer sites than machine planting, but results in less uniform planting quality (Long, 1991).

Whether planting by hand or mechanized planters, planting at the correct depth is essential. A relatively safe approach is to plant seedlings at the same level as they were in the nursery, container, or their natural setting (Stoddard & Stoddard, 1987). Survival of some species is reduced when planted as little as 1 cm too deep or too shallow (Weber, 1986). In most instances, the root collar should be even with the soil surface since the first small roots arise just under the collar. The root collar is usually at the soil surface of container-grown seedlings, but is more difficult to find with bare-rooted stock. Calibrated planting

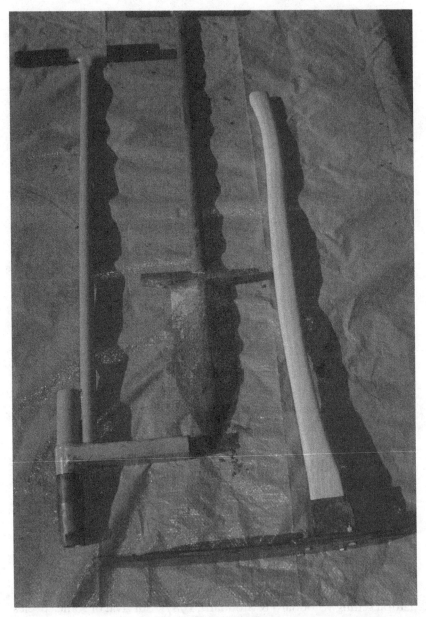

Figure 7.5. Hand implements designed for transplanting seedlings into wildlands. Seeding dibbles (left) remove a soil core for the seedling to be planted into. KBC bars (middle) are most appropriate in rocky soils that are difficult to penetrate. Hoedads (right) are used to create a small planting pit.

equipment and well-trained hand-planting crews are important aspects of success.

Hand planting allows additional care that increases seedling establishment. The seedling can be planted into the hole at the proper depth and the hole carefully filled, since shallow, loosely planted seedlings are more susceptible to early mortality. Moist soil from the bottom of the hole can be placed around the roots to create a basin to retain water on the surface. These basins are very effective in arid or semiarid ecosystems (Von Carlowitz &Wolf, 1991; Kavia & Harsh, 1993; Ffolliott *et al.*, 1994; Whisenant *et al.*, 1995). Litter or forest floor duff placed around seedling bases increases establishment. Seedlings with less exposure to sun or wind suffer less mortality (Long, 1991; Girard *et al.*, 1997).

Planting densities for trees and shrubs

The most effective planting density is a function of planting cost, availability of capital, growth rate, plant size at maturity, and management objectives (Stoddard & Stoddard, 1987; Le Houérou, 1992). It varies with species, climate and site quality. Erosion-control plantings are more effective when densely planted. Arid sites require more widely spaced trees and the removal of competing vegetation. Where water harvesting is used to increase seedling survival and growth, tree density (water harvesting area per tree) determines the amount of water that may be collected for each tree (FAO, 1989).

Larger *Atriplex* species reach maximum productivity at very high densities (10 000 ha^{-1}), but the plants are small and have increased mortality rates (Le Houérou, 1992). In contrast, more widely spaced plantings yield larger, more vigorous plants. Recommended planting densities for *Atriplex canescens* or *Atriplex nummularia* vary from 1500 to 3000 ha^{-1} in North Africa (Le Houérou, 1992) and Israel (Benjamin *et al.*, 1988). North African fodder plantations often grow *Atriplex* in rows spaced 4–6 m apart with individual plants planted 1–2 m apart within each row (Le Houérou, 1992). Smaller species require twice as many plants. Prostrate, creeping *Atriplex* species perform well when planted at densities as low as 1000 ha^{-1} because they expand to cover several meters (Le Houérou, 1992).

Trees that are not grown to maturity (i.e., Christmas trees, fuel wood, fodder) may be planted as close as 1.5 m apart in rows 1.5 m apart, whereas trees planted for timber production have other considerations. Trees grown on 2×2 m centers lose their side branches and produce more clear, knot-free wood (Stoddard & Stoddard, 1987). Tree spacing should seldom be less than 3×3 m (FAO, 1989). Trees planted at wider spacings retain their branches longer and produce wood with more knots. Closely spaced trees use available resources more effectively, are smaller, and produce more wood in less time. Closer spacings are most effective where maximum biomass production is desired (e.g., fuelwood plantation) (Ffolliott et al., 1994).

Wildings

Hand-held equipment or specialized equipment is used to transfer wildings from their natural settings to the repair site. Individual plants, root pads, sod and even large, multiple-plant soil transplants are used. Most of the foliage is trimmed off before planting because a large percentage of the root volume is lost (Munshower, 1994). This reduces transpiration losses and increases establishment success.

Root pads of thicket-forming shrubs with rhizomes or root sprouts establish with a high success rate. Front-end loaders remove soil pads, with multiple root segments of woody plants, from undisturbed sites (Munshower, 1994). These pads should be moved into similar-sized holes on the repair site and the air pockets filled with sand or other loose soil.

Sod

The top 3–8 cm of sod should be removed with hand tools or mechanical sod cutters and then placed on a well-prepared, moist soil and pressed firmly to insure contact between severed roots and soil. Deeply rooted bunchgrasses produce a sod that is difficult to handle. Certain situations may require the use of much larger sections of plants and soil. In Germany, surface area sections of sod 50 cm deep $\times$ 90 cm $\times$ 130 cm were moved prior to construction projects that would have destroyed sensitive areas (Bruns, 1988). While this approach is very expensive, it is an extremely effective way to salvage high priority areas.

Bare-root stock

Bare-root stock is widely used to establish woody plants (Munshower, 1994). Seedlings are grown in nurseries, lifted when dormant, and stored in cool, dark, and moist environments until transplanted. Once moved to the field and warmed, they become physiologically active and must be planted immediately (Munshower, 1994). The roots of bare-root seedlings should be kept covered until planted, because they are susceptible to death loss when the roots are exposed to sun or wind (Ffolliott *et al.*, 1994). One study found that 90% of Scotch pine (*Pinus sylvestris*) bare-root seedlings died after a 10-minute exposure to the atmosphere on a clear, hot day (Laursen & Hunter, 1986). Seedlings must be protected if planting is delayed once they are moved outdoors. Placing the seedling bundles in trenches and covering the roots with wet soil, peat, or sandy soil provides the best protection. Bare-root seedlings survive several days if protected this way (Ffolliott *et al.*, 1994).

Hardening induces dormancy by exposing the plant to ambient conditions similar to the planting site during transplanting (Munshower, 1994), such as reduced moisture, reduced nutrients, temperature extremes, and increased wind. This process reduces growth, accumulates carbohydrates, and makes the seedlings more tolerant of harsh conditions. It is induced in stages and usually takes 6 to 8 weeks (Ffolliott *et al.*, 1994). Hardened plants are more likely to survive the stresses associated with handling and transplanting (Weber, 1986). Unless properly hardened, bare-root stock has lower survival rates (Ffolliott *et al.*, 1994). Unlike bare-root stock, container-grown stock may be hardened or planted without hardening (Munshower, 1994).

Carefully planted bare-root stock is more likely to survive. Recommended procedures for hand planting bare-root stock were suggested by Ffolliott *et al.* (1994):

1. Clear planting spots of litter and scrape back the dry, surface soil.
2. Dig planting pit deep enough to avoid coiling roots.
3. Put plant against vertical rear (upslope) wall of pit at correct planting depth.
4. Fill the planting pit by holding the plant, filling pit, and firming moist soil around roots (firm bottom half carefully by hand and

top half with feet), add supplemental water if necessary and possible.

5. After planting, place loose, dry soil and litter around base of the plant.

Container-grown stock

Container-grown seedlings should be planted when dormant to increase establishment. This usually occurs when they are 6 to 7 months old. Once planted, they develop roots more quickly than bare-root stock. Transplanting is successful in both fall and spring, if exposure to harsh conditions is limited. A general rule for deciding when a seedling is the right size for transplanting is that the aboveground growth of container-grown stock should not be less than 0.2 m and no more than 1.0 m tall (Weber, 1986). The bottoms of peat pots are removed before planting to encourage early root expansion. Seedlings grown in reusable containers often develop malformed roots (i.e., root spiraling), which reduces root expansion when planted. It is advisable to remove the bottom (0.5 to 1.0 cm) of the root plug if the seedling roots are spiraled (Lefroy et al., 1991), by making two or three vertical incisions to a 1 cm depth in the root plug (Ffolliott et al., 1994).

Container-grown seedlings should be planted in pits large enough to accommodate the container or planting plug. They should be planted when the soil is moist and the soil firmed around the root plug or container to eliminate air pockets. Sufficient soil moisture contents are more easily achieved, in arid environments, if water collection pits, furrows, or microcatchments are installed before planting. Providing supplemental water significantly improves establishment rates; even single waterings are helpful (Ffolliott et al., 1994).

Cuttings

WOODY CUTTINGS

Poplars (*Populus* spp.) and willows (*Salix* spp.) are often transplanted as cuttings and effective techniques are available (Monsen, 1983; Morgenson, 1991). Cut poplar and willow whips 15–30 cm above ground level after natural defoliation and tissue hardening; then store them at -2 to -8 °C. These temperatures prevent fungal growth and premature bud opening. Wait until spring and stick the cuttings into a weed-free seedbed. It is also possible to immediately plant woody cuttings in some situations. In northern Shaanxi Province, China, 1-m willow stem sections were harvested in the late fall and pushed immediately into sand dunes with only 5–10 cm remaining above the surface (see Figure 5.4). These dune soils contained enough water for the cuttings to root and establish. Other studies recommend burying the entire cutting, with the top buried 1–3 cm, to reduce water loss from the exposed cut (Morgenson, 1991).

Soft stem sections can be rooted, grown in pots, and then transplanted. The cut ends of soft stem sections (current year's growth) are dipped into commercially available rooting hormones. The cuttings are then placed in a potting medium and kept under intermittent mist (Evans, 1991). Rooting is less successful with hardened woody stems. While too labor intensive for most wildland situations, rooted stem sections are useful for special situations. This technology facilitates the increase of genotypes with traits considered desirable for specific repair sites.

SPRIGS

Especially made sprig diggers or spring-tooth harrows harvest stolon or rhizome sections. Disking freshly cut hay into moist soil creates new plants of these species. Stems are less tolerant of careless handling than rhizomes. One rhizome sprig every 60–90 cm creates dense stands. Sprig establishment can be improved by (1) planting in moist soil; (2) using live sprigs soon after harvest; (3) planting sprigs deep and firming soil to reduce desiccation; and (4) creating a weed free environment (Burton & Hanna, 1985).

Maintenance of planted landscapes

At this stage, it is useful to reconsider a few previous points, reassess the situation relative to degree of degradation (Table 1.1), and discuss how the degree of degradation affects post-planting management. We should not forget that returning to the same management that caused the original degradation would degrade the site again. The benefits of many seedings are short lived, primarily due to poor species selection and inappropriate post-repair management. Degraded wildlands are often dominated by less desirable plants that prevent or reduce recovery success. Failure to remove these potential competitors may significantly reduce the available water, nutrients, and sunlight for seeded species during the critical early phases of establishment.

Plantings should be assessed when plants have had sufficient time to establish (Roundy & Call, 1988). Although several indicators of relative success have been proposed, the more common include: (1) number of plants per unit area; (2) distribution of plants; (3) apparent vigor; (4) stage of plant development; and (5) production (Vallentine, 1989). Damaged primary processes (hydrology, nutrient cycling, and energy capture) should be reexamined as part of the post-planting evaluation.

Prairies

The visual impact of most prairie plantings is disappointing after the first growing season, because initial perennial plant growth is predominantly belowground. Many seedings, considered failures after the first growing season, develop as planned during the second, third, and fourth years. Even with effective seedbed and planting procedures, three growing seasons may lapse before a recognizable prairie exists. Therefore, assessment of perennial plant establishment, particularly in difficult environments, should begin near the end of the second growing season. With effective seedbed preparation strategies that reduce early weed competition, most prairie plants are reasonably tolerant of competitive environments. Therefore, while mowing or herbicide applications may not be necessary during the first two growing seasons, they can accelerate prairie development (Thompson, 1992).

Since many prairie species do not develop tall shoots until the second or third growing season, it is possible to mow annual weeds without damaging young seeded plants. Mowing above the height of the seeded plants (15–25 cm) (Thompson, 1992) is most beneficial when the weed canopy shades 50–70% of the surface (Shirley, 1994). Herbicides must be used with great care. Do not broadcast herbicides to areas planted with forbs, shrubs, or trees unless you are sure they will not damage desired species. Although spot herbicide applications are useful for small sites, they are difficult to apply to large areas.

After three or four growing seasons, the site may have accumulated enough organic matter to consider burning. However, the presence of organic matter does not necessarily mean fire is appropriate. Burning is a long-term maintenance tool, rather than a prairie development tool, because it is most effective at reducing woody plant encroachment. Soil erosion is a potential hazard following burning, particularly on slopes. It is best, therefore, to burn early in the growing season, with wet soil to reduce erosion risks. The area should regrow soon, thus reducing the time when the soil is bare and unprotected. After five or six growing seasons, weeds should present few serious problems, although you may wish to overseed, hand weed, or use spot herbicide applications on small areas.

Forests and woodlands

Repair sites covered with herbaceous vegetation are particularly difficult environments for developing woody seedlings. Sites with a thick growth of herbaceous vegetation usually require weed control for the first 3 to 5 years, or until the desired overstory trees are taller than the competing plants (Thompson, 1992). Weeds may only be detrimental if the tree seedlings no longer receive sunlight. Weeds can be controlled with cultivators, harrows, mowing, herbicides, and even heavy mulching. Soil-disturbing mechanical approaches should be applied with caution on sites with high erosive potential. A 25-cm buffer around each seedling prevents physical damage to the seedlings. Restricting cultivation to the surface 6–8 cm reduces tree root damage. Herbicidal weed control treatments require selective application methods and/or selective herbicides to prevent seedling damage. As the stand develops, the developing trees may require pruning, training, or thinning. Herbivore damage to seedlings reduces seedling establishment in many situations.

Protecting shrub and tree seedlings

Recently transplanted seedlings are very susceptible to damage by extreme temperatures, high winds, moisture stress, and grazing pressure. Protection from these forces is often the deciding factor in successful establishment (Whisenant, Ueckert & Huston, 1985;

Bainbridge *et al.*, 1995). Numerous techniques protect plants from these problems. Tree shelters, plant collars, plastic mesh tubes, corrugated plastic tubes, and animal repellents are effective when applied to the proper situation.

CHEMICAL REPELLENTS

Recently transplanted seedlings are the most palatable plants on degraded landscapes. They attract rodents, rabbits, insects, deer, and livestock. Studies comparing the effectiveness of chemical repellents in protecting planted seedlings show variable results. Repellents are effective in some situations and less effective in others. Chemical repellents to discourage birds and rodents from eating seeds and seedlings greatly increased the survival of seeded trees (Stoddard & Stoddard, 1987). Without protection from seed predators, direct seeding pine and Douglas fir was risky due to the heavy loss of seedlings. Although fecal odors of predators and urine from coyotes (*Canis latrans*) significantly reduced browsing by mule deer (*Odocoileus hemionus*) on woody plants (Sullivan, Nordstrom & Sullivan, 1985), their use is impractical until synthetic scents become available. Soap bars eliminated (Scanlon, Byers & Moss, 1987) or reduced (Swihart & Conover, 1990) white-tailed deer (*Odocoileus virginiana*) browsing on apple trees and reduced browsing on Japanese yews (*Taxus cuspidata*) (Swihart & Conover, 1990). However, a comparison of chicken eggs, coyote urine, thiram, soap, and three commercial products concluded that if mule deer are moderately hungry, repellents will not deter browsing and fencing should be considered (Andelt, Burnham & Manning, 1991). Thus, the relative effectiveness of repellents for deterring deer depends on the hunger level of deer, the relative palatability of the species to be protected, and the concentration of repellent on the treated vegetation.

PROTECTIVE TUBES

Protective tubes reduce seedling damage by browsing herbivores (Figure 7.6). Solid tubes not only reduce browsing, they improve microenvironmental conditions around the seedling. They are constructed in several ways, but are most often solid-walled, corrugated plastic cylinders of variable height and diameter. Solid tubes are effective in arid ecosystems because they reduce light, decrease wind,

Figure 7.6. Protective tubes and nets used to increase the establishment of transplanted seedlings by reducing browsing damage by herbivores. The solid tubes also moderate microenvironmental conditions for developing seedlings.

and increase relative humidity around the seedling (Bainbridge *et al.*, 1995). Protective tubes made from plastic netting protect against browsing, but have little influence on microenvironmental conditions.

One-seed juniper seedling establishment and growth were significantly increased on northern New Mexico mine spoils when planted in either plastic mesh tubes or solid, corrugated plastic tubes (Fisher *et al.*, 1990). In another study, survival of white ash (*Fraxinus americana*), sycamore (*Platanus occidentalis*), and black walnut (*Juglans nigra*) seedlings was greatest (61%) in solid tubes, intermediate (43%) in nets, and lowest (28%) with no protection (Kost *et al.*, 1997). Plastic mesh tubes (with an inside diameter of 5 cm) effectively protect Douglas fir seedlings from herbivores where (1) severe damage is expected; (2) damage occurs during all seasons or by several wildlife species; or (3) restocking is necessary due to previous losses (Campbell & Evans, 1975).

8

Planning repair programs for wildland landscapes

Introduction

After assessing the status of each site, developing alternative techniques for repairing disrupted processes and redirecting vegetation change, we should reassess our objectives and develop the overall landscape repair plan (review Figure 1.5). The economic restrictions of most wildlands require that we minimize the energy expended by management actions. Therefore, our objectives should include fully functional wildland landscapes that are self-repairing, and have strong autogenic capabilities with minimal requirements for continuing subsidies. Repair programs, like conservation management programs, should be diverse, adaptive, self-organizing, and accept the ecological realities of change (Lister, 1998).

In addition to being fully functional and self-repairing, wildland landscapes must provide some combination of goods and services. These multiple objectives require carefully considered landscape designs, an organized planning approach, and monitoring procedures that measure progress and identify problems as soon as practical. This requires an understanding of ecological interactions in landscapes and practical guidelines for designing functional landscapes. Large, complex projects will require systematic decision-making processes that effectively organize and compare relevant information.

Table 8.1. *Site-specific problems and indicators of landscape-scale interactions with the potential to disrupt repair activities*

Indicators of site-specific problems	Indicators of potentially damaging landscape-scale interactions
Reduced vegetative and litter cover	Gully cutting (upslope or downslope
Deterioration of soil structure	from site)
(surface crusting, compaction, low	Excessive soil deposition
macro-porosity, reduced aggregate	Altered water table (higher, lower, or
stability, reduced infiltration)	reduced quality) suggesting disrupted
Reduced soil organic matter	hydrologic processes (transpiration,
Reduced water holding capacity	salinity, evaporation)
Wind or water erosion	Increased salinity due to run-on of low
Nutrient depletion	quality water
Reduced nutrient retention capacity	Low number and diversity of seed
Low functional and species diversity	immigrants
Depleted seed bank diversity	Accelerated nutrient losses to adjacent
Reduced activity and diversity of soil	landscape elements (fluvial, eolian, or
organisms	subsurface processes)
Soil salinity elevated above natural	Excessive animal damage (physical
conditions	damage, herbivory or seed predation)
	Inadequate pollination resulting in poor
	seed set
	Reduced landscape diversity
	Highly fragmented landscape

Source: Adapted and modified from Whisenant (1993).

Understanding landscapes

Structural and functional interactions of wildland landscapes affect their development and capacity for self-repair. Thus, failure to view repair sites as integral components of a larger, highly interconnected landscape can create inherently unstable landscapes. Not only must we assess and repair site-specific functional damage (Chapters 2 and 3) we must address landscape-scale interactions. This is our most difficult challenge. Landscape-scale problems are seldom obvious, but several potential problems are easily identified (Table 8.1). Although most indicators of landscape dysfunction are structural, we must look more closely for functional interactions. This requires a clear understanding of landscape structure and function.

Landscape structure

Landscape structure refers to the distribution, not the movement, of energy, materials, and species in relation to the sizes, shapes, numbers, kinds, and configurations of the landscape elements (Forman & Godron, 1986). Each part of a landscape differs in appearance from neighboring parts or from the surrounding matrix. Environmental differences and/or unique disturbance histories create these differences in the stature, age, or species combinations among each part of a landscape. Environmental resource patches result from differences in water availability, slope, aspect, soil type, or other environmental condition. Disturbances shaping landscapes may be either natural (fire, hurricane, wind blow-down) or human-caused (timber harvest, cultivation, altered hydrology, or abusive livestock grazing practices).

Landscape function

Landscape function refers to the flow of energy, materials, water, and species among the various parts of the landscape. Dysfunctional landscapes have altered resource regulation mechanisms, causing soil, nutrients, and water to be lost at unsustainable rates. These losses are especially relevant in wildlands, where the products and processes of damaged parts of the landscape can disrupt the stability of other parts. Since agricultural and degraded ecosystems have leaky nutrient cycles compared with undamaged wildland landscape elements (Allen & Hoekstra, 1992), their functional interactions with repair sites must be examined closely. Landscape structure influences function by affecting energy capture, hydrology, nutrient cycling, microenvironmental features, animal movements, propagule dispersal, and pollination processes. An appraisal of processes controlling the flows of limiting resources within and through the landscape help to identify the factors limiting recovery of that landscape.

Guidelines for designing landscapes

After reviewing the landscape-scale literature, Hobbs (1993) found little that was directly applicable to wildland repair programs. In fact, he believed little progress had been made since early theoretical guidelines for nature preserves (Diamond, 1975b; Wilson & Willis, 1975) that suggested bigger is better, and connected is preferred to fragmented. Since that review, numerous contributions provided additional guidance (Saunders & Hobbs, 1991; Aronson *et al.*, 1993b; Aronson *et al.*, 1993a; Aronson *et al.*, 1993c; Hobbs & Saunders, 1993; Hobbs *et al.*, 1993; Saunders, Hobbs & Erlich, 1993a; Whisenant, 1993; Bullock & Webb, 1995; Whisenant, 1995; Whisenant & Tongway, 1995; Aronson & Le Floc'h, 1996; Hobbs & Norton, 1996; Ludwig & Tongway, 1996; Tongway & Ludwig, 1996; Ludwig *et al.*, 1997). Despite those contributions, there is still a great deal to be learned.

We lack guidelines to quantify 'how big is big enough' or 'how connected should a landscape be' (Hobbs & Norton, 1996). Our understanding of how landscape structure influences function is limited and our ability to design fully functional landscapes is even more limited. It is probable that each set of site condition, site history, and repair goals require unique combinations of size and connectedness. Thus, step-by-step cookbook approaches are unlikely to have broad applicability. We must consider each situation independently, and design repair programs specifically for those circumstances. Fortunately, several guidelines have emerged that should be considered when attempting to design functional wildland landscapes (Table 8.2). Since landscape interactions were an underlying theme of all previous chapters, these guidelines reiterate those concepts. They provide a systematic structure for considering landscape interactions, but do not produce specific, quantitative designs.

Treat causes rather than symptoms of degradation

Treating the symptoms of degradation without addressing the causes is a recurring theme of failed wildland repair programs. Chronically overgrazed areas have been reseeded and then returned to the management regime that caused the original problems. The benefits from these repair efforts were short-lived.

Table 8.2. *Guidelines for designing functional wildland landscapes*

Although not all of these guidelines are appropriate for each situation, they should be carefully considered for adoption to the unique circumstances of each landscape.

Guideline	Comments	Additional relevant information
Repair goals		
Treat causes rather than symptoms of degradation	Repair strategies that do not address the original causes of damage are unlikely to achieve long-term goals.	Chapters 1, 2, 3, and 4 (Brown & Lugo, 1994; Milton *et al.*, 1994)
Emphasize process repair over structural replacement	The recovery and maintenance of processes, rather than species, is the key to ecosystem repair and resilience. This does not prevent the return of certain species, only that they are used to repair function.	Chapters 1, 2, 3, 4, and 5 (Breedlow *et al.*, 1988; Mitsch & Cronk, 1992; Mitsch & Gosselink, 1993; Brown & Lugo, 1994; Whisenant & Tongway, 1995; Wyant *et al.*, 1995; Bradshaw, 1996; Hobbs & Norton, 1996; Ludwig & Tongway, 1996; Tongway & Ludwig, 1996; Tongway & Ludwig, 1997b)
	Self-repair and self-design are initiated with designs that use solar energies to increase the capture and retention of limiting resources. This reduces maintenance requirements.	
Allow sufficient time for autogenic processes to operate	Autogenic processes are desirable because they operate without cost, are self-sustaining, and may be applied on a large scale. However, they operate slowly in some environments.	Chapters 3, 4, and 7 (Mitsch & Cronk, 1992; Mitsch & Gosselink, 1993; Bradshaw, 1996)
Repair programs should address the largest scale at which the process damage occurred	Processes damaged at one scale influence all smaller scales and large-scale problems are not adequately assessed or repaired at much smaller scales.	Chapters 1, 2, 3, 4, and 8 (Saunders *et al.*, 1993a; Lewis *et al.*, 1996; Rabeni & Sowa, 1996)

Site stabilization and the repair of primary processes

Design landscape to increase retention of limiting resources	Take full advantage of landform features with inherently greater resource retention capabilities.	Chapters 1, 2, 3, 5, and 6 (Whisenant & Tongway, 1995; Hobbs & Norton, 1996; Ludwig & Tongway, 1996; Tongway & Ludwig, 1996; Tongway & Ludwig, 1997b)
	Increase biotic controls over limiting resources where necessary.	
Incorporate spatial variation (of methods, landscape patches, planting dates, and/or species mixtures) into the landscape design as a 'bet-hedging' strategy	'Bet-hedging' limits the damage if strategy for a particular landscape component fails, since an adjacent area may be successful and expand.	Chapter 8 (Pastorok et al., 1997)
	Multiple strategies contribute additional information for future adaptive management efforts.	
Design landscapes to maintain the integrity of primary processes	Restore or maintain the integrity of processes supporting proper functioning of hydrology, nutrient cycling, and energy capture.	Chapters 1, 2, 3, 4, 5, and 8 (Haila et al., 1993; Saunders et al., 1993b; Saunders et al., 1993a; Whisenant & Tongway, 1995; Hobbs & Norton, 1996; Ludwig & Tongway, 1996)
	Protect and enhance remnant patches where possible.	
Design linkages into the landscape, taking landscape patchiness into account	While corridors may serve as a conduit for some species, they act as a barrier to the dispersal of others.	Chapters 3, 4, and 8 (Saunders et al., 1993b; Saunders et al., 1993a; Hobbs & Norton, 1996)

Table 8.2. (cont.)

Guideline	Comments	Additional relevant information
Propagule dispersal, animal interactions and pollination requirements		
Design propagule donor patches into the landscape	Propagule donor patches provide a continuing supply of propagules to the surrounding landscape.	Chapters 4, 5, 6, 7, and 8 (Robinson & Handel, 1983; Janzen, 1988b; McClanahan & Wolfe, 1993; Whisenant, 1993; Debussche & Isenmann, 1994; Kollmann & Schill, 1996; Lamb *et al.*, 1997; Parrotta *et al.*, 1997)
Encourage animal dispersal of desired seed	Assess the potential for animal dispersal of problem-causing species and consider arranging landscape components to reduce dispersal of problem species. This includes the use of attractant patches, artificial perches, or the intentional feeding of desired seed.	
Design landscape to encourage wind dispersal of desired seed	Propagule donor patches might be strategically located to increase the effect of wind on the transport of desired species.	Chapters 4, 5, and 6 (Jackson, 1992; Whisenant, 1993; Greene & Johnson, 1996; Schwarzenbach, 1996; Timoney & Peterson, 1996; Hodkinson & Thompson, 1997)
	Assess the potential for wind dispersal of problem-causing species and consider arranging landscape components to reduce their dispersal by wind.	
Design landscape to encourage positive animal interactions while reducing detrimental animal interactions	Earthworms, termites, ants, beetles, and certain vertebrates improve mixing, aeration, and friability through their burrowing activities. Herbivory and seed predation can seriously reduce establishment success.	Chapters 4, 5, and 6 (Humphreys, 1981; Crawley, 1983; Majer, 1989; Archer & Pyke, 1991)

Design landscape to meet pollination requirements of critical species	Problems with animal-pollinated species are most frequent when the repair site is small, isolated from vegetation, and populated by species not found in the adjacent vegetation.	Chapters 4, 5, and 6 (Majer, 1989; Menges, 1991)
Microenvironmental modification		
Use species that improve microenvironmental conditions on a local scale	Select species for their ability to ameliorate harsh microenvironmental conditions.	Chapters 3, 4, and 5 (Allen & MacMahon, 1985; Farrell, 1990; Brooks et al., 1991; Satterlund & Adams, 1992; Vetaas, 1992; Whisenant et al., 1995)
Arrange landscape components to improve microenvironmental conditions on a larger scale	Include shelterbelts or patches of taller vegetation to affect wind and solar regimes of larger areas.	Chapters 3, 4, 5, 7, and 8 (Bird et al., 1992; Ryszkowski, 1995; Mohammed et al., 1996; Grant & Nickling, 1998)

We must start by evaluating the effect of current and previous management. Why are these processes damaged? Disturbances such as deforestation, cultivation, or mining are usually obvious, but other disturbances may be less obvious. Altered fire regimes (either too frequent or too infrequent) also cause significant degradation. Are the damaged processes the result of management actions that occurred on some other part of the landscape?

Understanding the causes of damage simplifies the design of repair strategies and increases the likelihood of success. Those causes that have only damaged the biotic component are often repaired with improved management alone. However, damaged primary processes require repair of the physical system (e.g., soil surface or microenvironmental attributes) rather than only working with the biotic system (adding seed or plants). It is far more difficult to reverse the impacts of actions that degrade the resource base and the ability to capture resources (Brown & Lugo, 1994; Milton *et al.*, 1994). Strategies that direct vegetation change (Chapter 4) or add plants to the system (Chapters 5, 6, and 7) are effective where primary processes have not been significantly damaged (e.g., Table 1.1).

Emphasize process repair over structural replacement

A summary of the self-organizing ability of ecosystems concluded that managers should cultivate the capacity of natural systems for self-organization rather than trying to control them (Hollick, 1993). Therefore, repair strategies that view nature as a flexible and adaptive partner rather than an adversary have the most promise for the low-input situations most prevalent on wildlands.

Wildland repair programs have traditionally focused on replacing species or nutrients (structure). However, it is now widely understood that the maintenance of processes, rather than structure, is the key to ecosystem resilience and repair. This book endorses that philosophy by emphasizing the repair of damaged primary processes at multiple spatial scales. This philosophy requires strategies that direct vegetation toward general objectives, rather than narrow, predefined objectives.

Design repair actions at the proper scale

Many repair projects failed because problems and solutions occurred at different spatial or temporal scales. Large repair programs must identify appropriate scales for assessing and addressing problems. Since ecosystem processes and structures show properties at multiple scales, the scale of our observations is critical (Lewis *et al.*, 1996). Processes damaged at one scale influence all smaller scales (Lewis *et al.*, 1996) and large-scale problems cannot be adequately assessed or repaired with data or actions limited to much smaller scales. Therefore, repair programs should address the largest scale at which the stress (e.g., process damage) occurs (Rabeni & Sowa, 1996). Regardless of what scale we focus on, we must examine larger scales for an understanding of context and smaller scales for insight into underlying mechanisms (Lewis *et al.*, 1996). Although repair actions may be required at regional or national levels, local actions are often the only practical option (Lewis *et al.*, 1996). It is often necessary to begin at smaller scales, while we work towards resolution of the problem at larger scales.

Design landscapes to increase retention of limiting resources

Where possible, watershed boundaries are the most appropriate boundaries for management (Oyebande & Ayoade, 1986; Thurow & Juo, 1991; Korte & Kearl, 1993; Thurow & Juo, 1995) and for manipulating hydrologic and geochemical processes (Thurow & Juo, 1991). Since ecosystems are open and fences, political jurisdictions, or ownership boundaries do not restrict primary processes, landscape-scale repair efforts that operate on watershed boundaries are desirable. Although we will likely be limited to working on specific sites, rather than entire watersheds or large landscapes, we can still address many of the landscape-scale problems associated with damaged wildlands. Limiting resources are controlled by landform attributes or biotic controls (Chapter 2). We must understand the relative impact of both, across the landscape, to increase resource retention in the landscape.

First, we must consider the relative ability of each landform type to capture, filter, and retain soil, water, nutrients, and organic materials. Resource retention mechanisms operating across landscapes become clear following a geomorphic analysis of the relative position of the site with the larger landscape (see Figure 2.1). This identifies sites that either lose or gain water, soil, and nutrients through runoff or deposition. Site position, within the landscape, suggests the nature and magnitude of fluvial processes. Each part of the landscape has very different obstacles and potentials, largely determined by their relative position. This relative position affects the rate of runoff, amount of water captured (e.g., closed basins), erosion potential, and potential for capturing resources from other parts of the landscape. For example, concave sites on an otherwise flat landscape receive relatively less input from adjacent sites than similar concave sites located at the base of a long slope.

Second, we must determine the influence of biotic controls over the flows of water, nutrients, and organic materials. Biotic controls increase with more resource control patches (fine-grained), wider patches, and closer patches (Ludwig & Tongway, 1995). A landscape function analysis assesses the magnitude of each of these biotic control attributes. Landscape function analysis involves collecting vegetation and soil surface information from contiguous quadrats positioned along an environmental gradient (i.e., slope or wind direction). Comparing foliar and litter ground cover and the upslope (or upwind) distance between obstructions on damaged and relatively damaged sites gives a relative assessment of resource regulatory mechanisms on each site. This provides insight into the most appropriate resource regulation mechanisms and the scale at which they operate. These data can be analyzed with boundary analysis (Ludwig & Cornelius, 1987) to quantify patchiness and identify patches at multiple scales (500 m for landscape scale or 1 m for soil surface features) (Ludwig & Tongway, 1995).

Some problems with landscape function are attributable to the breakdown in fine-scaled patchiness or the fragmentation of landscapes (Schlesinger *et al.*, 1990; Ludwig & Tongway, 1995; Schlesinger *et al.*, 1996; Huston, 1997). Although resource conservation is generally greater in landscapes dominated by fine-scale biotic regulation, patchy resource distributions may be desirable where resources are very limited. For example, in arid regions a patchy distribution of resources (e.g., water) is more productive than the same amount of resources uniformly distributed over the landscape (Noy-Meir, 1973). This occurs because disturbances such as cultivation or extreme grazing practices homogenize landscapes – uniformly reducing water availability, soil nutrients, and organic matter. Thus, degradation that leads to spatial leveling of resource creates landscapes where limiting resources are uniformly below the establishment threshold for the desired plant species (Whisenant, 1993). In this uniformly limited landscape, actions that concentrate resources allow patches of vegetation to develop. This is desirable, although it depletes some of the landscape of resources, because it facilitates the establishment of vegetative patches where none existed before (Whisenant, 1993; Tongway & Ludwig, 1996; Herrick, Havstad & Coffin, 1997; Tongway & Ludwig, 1997a; Tongway & Ludwig, 1997b). This resource distribution has been termed the 'Robin Hood in reverse' effect because it robs the poor to pay the rich (Tongway & Ludwig, 1997a). Autogenic processes may subsequently expand the size of vegetation patches and associated resources. Thus, in severely depleted landscapes, strategies that capture resource flows contribute to resource retention and autogenic development, although they create a more patchy landscape.

Not all patches leak resources equally, with nutrient losses tending to increase with patch size and patch duration (Ewel, 1997). Resource flows can be manipulated to help achieve repair objectives, since landscapes rich in ecotones probably lose fewer nutrients (Ryszkowski, 1992). Small vegetation gaps lose fewer nutrients, because roots from the surrounding plants extend under much of the gap. Short-lived gaps retain nutrients in the organic component (both aboveground and belowground). Nutrient losses are greater when patch size exceeds the reach of roots from the surrounding plants and in old gaps where the organic materials have decomposed (Ewel, 1997). Nutrient losses are typically highest where rainfall greatly exceeds evapotranspiration. Where evapotranspiration exceeds precipitation and fully functional ecosystems contain some bare ground, nutrient losses (through leaching or overland flows) from gaps may be limited to unusually wet events or high rainfall systems (Ewel, 1997). Vegetation gaps are subject to significant nutrient losses through wind erosion.

Design spatial variation into landscapes

The size and shape of boundaries between landscape patches affect the rate and direction of successional processes. In the absence of complete knowledge, we seek spatial diversity. Size is important because small patches seem to be more susceptible to invasions by unwanted species than larger areas (Ewel, 1997). Interior species require large patches, with little edge effect. The effect of boundary shape is illustrated by colonization rates from mineland sites in New Jersey (Hardt & Forman, 1989). Sites adjacent to concave forest boundaries had 2.5 times more colonizing stems than those near convex boundaries did. Colonizing stems established over 61 m from concave boundaries, but rarely grew further than 13 m from convex boundaries (Hardt & Forman, 1989). Stem densities of animal-dispersed species compared to abundance in the adjacent forest edge, whereas no relationship existed for wind-dispersed species. Colonization patterns near straight boundaries were intermediate between those opposite concave and convex boundaries.

Design landscapes to maintain the integrity of primary processes

Patches of disturbed natural vegetation or cultivated fields are leaky, because they retain little of their annual nutrient input (Allen & Hoekstra, 1992). The functioning of small, remnant patches, within greatly modified landscapes, becomes dominated by their context (Hansen, Risser & Castri, 1992; Haila *et al.*, 1993). Landscapes that maintain the integrity of primary processes can reduce the impact of these problems. Carefully designed landscapes retain limiting resources, even if cultivated fields or disturbed patches remain.

Plant structure and the arrangement of plant structure influence the partitioning of solar energy (Ryszkowski, 1989), that affects the hydrology and nutrient retention over the larger landscape (Burel *et al.*, 1993; Hobbs, 1993). Trees transpire more water than meadows or cultivated fields. This alters the chemical composition of ground water within direct and indirect (capillary ascent) root range (Ryszkowski, 1992). In a Polish landscape of agricultural fields and riparian forests, croplands released most of the nitrogen input and much of the phosphorus input (Bartoszewica & Ryszkowski, 1989). However, adjacent forests released only about 10% of the annual N inputs and 20% of the annual P input. The pathways of nutrient losses from the agricultural fields to the discharge stream were predominantly surface flows for P and subsurface flows for N. Had the riparian forest not been present, the discharge stream would have received twice as much N. Most, 60–75%, of all nutrients were captured within the lateral flow through the first 19 m of the riparian forest. Thus, the shelterbelts and remnants of natural forests controlled nutrient fluxes and increased the nutrient holding capacity of the entire landscape.

Design linkages into landscapes

How connected or fragmented should the repaired landscape be? There is no single answer, but it depends on the spatial distribution of patches and on the scale at which organisms interact with landscape pattern (Keitt, Urban & Milne, 1997). Although a landscape may 'appear' connected to well-dispersed species, it may 'appear' excessively fragmented to poorly dispersed species. Thus, landscape structure is a scale-dependent filter that differentially influences species movements. Landscapes with gradual boundaries have a more subtle filtering influence compared with highly fragmented landscapes with abrupt changes. Protecting remnant patches with buffers and linking remnants to similar patches are often recommended, although little empirical support is available. Buffers reduce the negative impacts of external forces on small remnants (Haila et al., 1993; Saunders et al., 1993a). Corridors between similar vegetation facilitate species movements and the integrity of some ecosystem processes (Saunders et al., 1993a). However, few guidelines exist to design their use. Frequent disturbances and large edge to area ratios complicate corridor management. Corridors are dominated by external rather than internal influences, unless they are wide enough for an interior that is not influenced by edge effects (Hobbs, Saunders & Hussey, 1990). Since the critical attributes of corridor design are species specific, any single corridor is unlikely to be effective for all components of the biota (Hobbs et al., 1990). Although corridors serve as conduits for some species, they are barriers to other species.

Design propagule donor patches into landscapes

Despite the vast areas that require assistance, economic, social, and political constraints make it unlikely that large-scale repair actions will occur on most wildlands. Therefore, it becomes especially important to develop strategies that take advantage of natural seed production and natural seed transport mechanisms (animals, wind, and water). Thus, we develop landscape-scale strategies to increase natural dispersal by (1) planting propagule donor patches; (2) increasing animal dispersal of propagules; and (3) increasing the effectiveness of wind as a natural transport mechanism.

The scale of damaged wildlands and the chronic shortage of resources for their repair suggests the value of landscape-scale strategies that construct landscapes with propagule donor patches (Whisenant, 1993). These donor patches continue to release propagules into adjacent landscape patches. This continuing supply of propagules will eventually increase recruitment into ecosystems with episodic seedling establishment. Where natural seedling establishment occurs only once in 5 or 10 years, the probability of success following artificial seeding is small. Creating many donor sites, distributed over the landscape, provides a continuing source of propagules (Whisenant, 1993). A variety of empirical and theoretical studies suggest that seed dispersal is more effective from many small satellite stands rather than from fewer large stands (Moody & Mack, 1988). Ultimately, grazing management (or other landuse practices) will determine the success of donor patch strategies.

Landscape designs that increase seed movement may accelerate successional changes on many parts of the landscape. Relying exclusively on natural processes to establish plant species can be slow and highly variable, depending on the climate, soils, and the availability of propagules. Often, the pool of desirable species is insufficient to initiate recovery, or the recovery period would be unacceptably long. For example, natural seed immigration into industrial waste heaps in northwest England produced an impoverished flora, even after 50 to 100 years (Ash et al., 1994). The distance to suitable seed sources (40 km) limited their development. Natural recruitment rates depended on land unit size, proximity to natural seed sources, intensity of previous cultivation, competitiveness of desirable plants, precipitation, grazing pressure, and the extent of soil erosion. As an example, heath vegetation in infertile, old fields in Dorset, England, reestablished without intervention when the landscape contained propagule donor sites or allowed long-distance transport of seed by animals (Smith et al., 1991). In the absence of features that increased immigration of desirable species, repairing the old fields required management strategies that included artificial seeding. In Canada's Wood Buffalo National Park, natural regeneration was ineffective when clear-cut size exceeded the natural dispersal distance of white spruce trees (Timoney & Peterson, 1996).

Design landscapes to encourage animal dispersal of desired seed

Patches of 'attractant plants' attract greater numbers of seed-dispersing animals to specific sites (Lamb *et al.*, 1997). Thus, planting 'attractant species' to attract seed-dispersing animals should result in the recruitment of more species than artificial planting alone. The attractant plants may provide food, shelter, or perching sites for the seed-dispersing animals. In the northeastern United States, large urban landfills seldom develop into diverse woodlands (Robinson & Handel, 1983). One year after establishing a plantation of 17 species (shrubs and trees) on a New York landfill, 95% of the new seedlings came from outside the plantation. Most (71%) of those were from fleshy-fruited, bird-dispersed plants common in nearby woodland fringes. Even though the species originally planted in the plantation had not yet begun to produce seedlings, the woody plants attracted birds that dispersed at least 20 new plant species to the site (Robinson & Handel, 1983). Locations with more trees than shrubs had proportionally more additions, suggesting the importance of plant size in attracting birds that import seed.

Highly mobile, fruit-eating birds are very effective at moving seed into wildland repair sites. Where mice may have maximum dispersal distances of 10–20 m (Kollmann & Schill, 1996), birds disperse seed at much greater distances. Repair strategies deliberately designed to attract birds for bringing seed into the area have considerable potential. Bird dispersal of fleshy-fruited plants is more important where the vegetation is heterogeneous. This often occurs where woody patches develop in open vegetation and where grassy patches appear in forests (Debussche & Isenmann, 1994). It is also common where forest patches are small, distances are great, and seed banks are depleted (McClanahan & Wolfe, 1993). Thus, encouraging birds to import seeds is most beneficial in highly fragmented landscapes.

Although strategies that rely on birds to bring in seed from the surrounding landscape are attractive, they do not always yield the desired results. For example, ten years after establishing tree plantations into a former bauxite mine in Amazonia, new species recruitment was limited to smaller-seeded forest trees (Parrotta *et al.*, 1997). Fruit-eating birds and mammals on the site were limited to species that feed on small-seeded species. The birds and mammals that typically disperse the larger seeded plant species were still rare in the reforestation area. Although the reforestation program successfully recreated favorable environments for the regeneration of native species, limited seed immigration restricted the recruitment of many important species (Parrotta *et al.*, 1997). Thus, in this and in many other situations, we may need to bring in additional species or remove some of the species that come in naturally.

As part of a long-term restoration plan, livestock were used to distribute seed through 700 km² of dry tropical forest in northwestern Costa Rica (Janzen, 1988a; Janzen, 1988b). Seed dispersal patterns of the livestock, combined with a diverse, native fauna, increased the plant diversity over large areas at little cost. Their resting, foraging, and feces deposition patterns concentrated defecated seed along ravines, rock outcrops, and under isolated trees. Isolated pasture trees play a particularly important role in dry forest expansion because they initiate expanding islands of animal-dispersed, woody vegetation. With time, these forest patches expand and coalesce with other patches. As a result, forest development is greatly accelerated by the presence of isolated trees – trees which are almost exclusively dispersed into large, open pastures by horses and cattle (Janzen, 1988a). Thus, livestock not only initiate the development of the original trees that accelerate subsequent forest development, they are necessary for seedling survival. Dense stands of 1–2 m tall grasses block sunlight, capture nutrients, and fuel fire, all factors that prevent dry forest development. Once the sites reach a stage where the grass is no longer seriously threatening the woody succession, the livestock are removed.

Design landscapes to encourage wind dispersal of desired seed

Although wind transports seed great distances, it does not selectively place them into safe sites. Wind-dispersed seed also contains undesirable species. Seed dispersal by wind is influenced by prevailing wind direction during the season of seed maturation, relative topographic position of donor and receptor sites, and presence of structures (trees, shrubs, rocks, etc.) that trap a disproportionate share of the seed blowing through the area. After considering the prevailing wind duration when seed dispersal occurs, we can strategically place stands of wind-dispersed species. Planting sites with higher topographic positions increase the dispersal distance of wind-dispersed seed. These scattered plants or patches produce seed year-after-year, increasing the odds of having seed available for those episodic establishment events that typify many wildlands.

Design landscapes to encourage positive animal interactions

Dynamic interactions among plants and animals are important regulators of ecosystem development and maintenance. Animals affect succession through herbivory, seed predation, seed dispersal, pollination, soil structure and turnover, and litter decomposition and nutrient cycling. Both vertebrates and invertebrates contribute to ecosystem structure and functioning and may have significant influences on repair efforts. The impact of animals on succession is profound and has been addressed in several books (Crawley, 1983; Majer, 1989) and review articles (Majer, 1989; Archer & Pyke, 1991; Jones et al., 1994; Pollock et al., 1995; Jones et al., 1997).

Wildland repair programs are improved with strategies that favor certain groups of animals while discouraging others (Archer & Pyke, 1991). Strategies that manage herbivory, encourage dissemination of desirable propagules, discourage predation of planted seed, and attract pollinators contribute toward repair success (see Chapter 4 for additional discussion). Our understanding of the impact of animals on natural succession and other types of vegetation change is inadequate to precisely define their potential role. However, current knowledge is sufficient to allow several generalizations (Whelan, 1989). Intensive herbivory reduces diversity and eliminates palatable species. Moderate herbivory increases species diversity. Size and structural characteristics of surrounding landscapes determine the amount of seed dispersed by animals. Pollination concerns become more important in well-developed sites where the surrounding landscape contains different species. Burrowing and trampling are important because they contribute to soil turnover and increase species diversity by creating colonization sites for additional species. The design of the repair site relative to its surroundings will influence each of the preceding generalizations. The shape, size, and arrangement of each repair patch and the surrounding vegetation influence seed dispersal, seed predation, herbivory, and pollination.

Design landscapes to improve the microenvironment at different scales

Harsh microenvironmental conditions can be improved at both local (individual plant) and landscape scales. The amelioration of harsh microenvironments by woody plants is a pervasive theme in ecology, ecological restoration, and agroforestry because it has significant management implications. For example, juvenile pines (*Pinus strobus* and *Pinus resinosa*) were found beneath oak (*Quercus rubra*) canopies at densities over six times that occurring in open areas of Ontario, Canada, but only when the oaks were at least 35 years of age (Kellman & Kading, 1992). This delayed effect suggests the importance of physical stature (e.g., larger trees have more impact). In arid and semiarid ecosystems, woody plants improve microenvironmental conditions by moderating wind and temperature patterns (Allen & MacMahon, 1985; Vetaas, 1992; Whisenant *et al.*, 1995).

Shelterbelts, which are patches of woody vegetation within a matrix of shorter vegetation, reduce evapotranspiration rates in a zone from 4 to 8 times the shelterbelt height. That zone is more water efficient and more productive. In the Amazon Basin, biomass production within 100 m of a forest edge was up to 36% lower than the interior of the forest (Laurance *et al.*, 1997). Microenvironmental changes along fragment edges increased tree mortality when fragments were smaller than 100–400 ha. Chapters 3, 4, 5, and 7 describe several strategies for ameliorating harsh microenvironments with individual plants or landscape patches.

A decision-making framework

After developing repair alternatives for each site, we need a systematic decision-making approach for developing the final plan. Planning ecological repair programs requires numerous decisions, involving large amounts of information. A planning and decision framework developed for ecological restoration programs (Wyant *et al.*, 1995), aids the development of repair programs with the capacity to produce the desired goods and services. A decision framework for environmental restoration (Pastorok *et al.*, 1997) addresses additional ecological concerns. This planning and decision framework involves context analysis, risk assessment, and the development of site-specific plans for

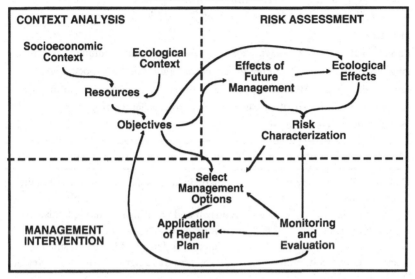

Fig. 8.1. A decision framework for selecting from among multiple repair options
during the planning process. After Wyant *et al.* (1995). Used with the permission
of Springer-Verlag New York, Inc (*Environmental Management*).

management intervention (Figure 8.1). Although decision frameworks
organize information, they neither explain what is happening nor
predict the results (Wyant *et al.*, 1995).

Context analysis (of the socioeconomic and ecological contexts) and
risk assessment (of climatic, technical, and socioeconomic uncertainties) help select from among alternative repair strategies to develop a
comprehensive landscape-scale repair plan with site-specific components. Long-term monitoring and evaluation programs must have
clear, achievable, site-specific goals.

Context analysis

The ecological and socioeconomic context involves numerous considerations (Table 8.3) for setting goals and planning wildland repair programs (Wyant *et al.*, 1995). The socioeconomic context includes
economic, esthetic, religious, and subsistence issues. The ecological
context includes structural and functional considerations. It refers to

Table 8.3. *Selected socioeconomic conditions and ecological features of the surrounding that influence the context for wildland repair sites and programs*

Large repair programs involving many people, large areas, multiple goals, and long time frames usually require formal consideration of the socioeconomic and ecological contexts. Although appraisals of the socioeconomic context may be simple and rapid for small, privately funded projects, larger projects require increasingly detailed assessments.

Socioeconomic conditions	Ecological features of surrounding area
Community involvement or interest in project	Climatic factors (solar radiation, wind regime, relative humidity, precipitation)
Land tenure issues	Nutrient and water fluxes among landscape
Community goals	components
Human population density	Presence and density of large and small
Local economic conditions	herbivores that might be attracted to repair
Regional economic conditions	site
National economic conditions	Presence of seed predators
Global economic conditions	Propagule movements (animal, wind, water
Government policies	carriers)
Demand for products and services from site	Fire regime of surrounding area (ignition sources, flammability of fuel, amount and
Knowledge of stakeholders	continuity of fuel)
Administrative barriers	Hydrologic conditions
Social, political, and economic instability	Relative topographic position
	Soil erosion (wind or water)
	Pollinator availability
	Diversity of species and landscape patches
	Quality of incoming water

the spatial influences on a system or the spatial connections of a site with its surrounding landscape.

SOCIOECONOMIC CONTEXT

Destructive management practices occur for many reasons. The causes of degradation, and the barriers to effective repair, must be identified and halted (Table 8.3). Repair strategies that only address the symptoms of the problem and fail to consider the underlying socio-politico-economic causes of degrading practices are doomed to failure. In many countries, population pressures or the land tenure system prevent lasting repair of damaged wildlands.

The decision process should assess the opinions of all people affected by the repair program (stakeholders). Small repair projects on private property may have few stakeholders, simple goals, and short time frames that require little structured planning. However, larger programs involving many people, large areas, multiple goals, and long time frames usually require more formal planning (Ffolliott et al., 1994). As repair projects become more complex, formal planning can assure effective use of labor and resources while meeting the project objectives.

The socioeconomic context includes community goals, traditional cost-benefit studies, land tenure issues, political issues, the development of alternative visions about the desired outcomes of the program, and administrative barriers to working at landscape levels. Whereas appraisals of the socioeconomic context may be simple and rapid for small privately funded projects, larger projects require increasingly detailed assessments. Technical assistance agencies and international financing organizations (e.g., World Bank and the Interamerican Development Bank) now require community involvement during project planning as a way to assess the socioeconomic context. Limited stakeholder knowledge about ecosystem services, the natural resource base, and ecological processes are a common problem (Wyant et al., 1995). The ability of local stakeholders to provide informed opinions on the ecological repair of damaged ecosystems depends on their knowledge of ecosystem function and their access to knowledge. Greater understanding of ecosystem processes can improve their ability to judge the potential benefits and characterize the risks associated with various repair alternatives. Therefore, social context analyses often include an appraisal of the stakeholders' understanding of relevant issues (Wyant et al., 1995). It may also require community education programs that develop an ecological awareness that allows them to make informed decisions about repair goals and strategies.

Conservation biologists were among the first to implement landscape-scale wildland repair programs based on the development of socioeconomic synergies (Janzen, 1988a; Janzen, 1988b). Both ecological and biocultural restoration of the dry tropical forests of Costa Rica are addressed with a program that encourages natural regeneration of degraded forest remnants and surrounding farmland. The success of this approach demands landscape and landuse patterns that address the needs and desires of local inhabitants. Achieving that objective requires that local inhabitants have a genuine interest in the success of the project.

ECOLOGICAL CONTEXT

Ecological context analyses remind us that degraded wildlands are embedded within dynamic, ever-changing landscapes that influence what occurs on repair sites (Table 8.3). The physical context of a site is physically larger and surrounds the site. The temporal context considers longer time periods and is more constant over time than present conditions. Context is situation- and species-specific. The physical context for hydrologic functioning in an estuary of a major river may contain millions of square kilometers. However, the operative context for an epiphyte within that estuary is much smaller.

Wildland repair sites have both inherent and induced site limitations. Inherent limitations are determined by climatic and geomorphic features (e.g., the substrate's chemical status influences the site's nutritional status) of the site. Induced site limitations are primarily associated with degradation (e.g., erosion, deforestation, overgrazing, secondary salinization, and other human-caused problems). As stated in previous chapters, assessing the ecological context of a site should consider (1) the abiotic environment; (2) nutrient cycling; (3) hydrology; and (4) community-level vegetation processes. Since ecological systems are open, their properties are not only determined by what happens within each system, but also by interactions with other parts of the landscape (Allen & Hoekstra, 1992; Pickett & Parker, 1994). Thus, it is important to consider the interactions of hydrology, seed movement, animal movement, energy flow, and nutrient transport among different parts of the landscape.

Effective management recognizes what is missing from the ecological context and replaces those missing services (Allen & Hoekstra, 1992) by providing subsidies that replace the services provided by the destroyed context. Attempts to restore the historical context to repair sites will always be constrained in some way. Small repair sites will have limitations imposed on them by their context. Fire regimes, nutrient fluxes, and river flows are but a few of the processes that may be constrained by their context. Fire may begin outside the site, but depends upon the relative continuity of fuel to allow it to move from site to site. Thus, the context of a site determines whether historical fire regimes continue. The context influences the flux of water and nutrients to downslope (or downwind) sites. The flow of propagules to a site is strongly influenced by its context (vegetation in adjacent landscape patches) and propagule transport mechanisms.

Risk and uncertainty

ASSESSING RISK AND UNCERTAINTY

Risk assessment (Figure 8.1) requires an understanding (or best estimate) of the uncertainties associated with all parts of the repair program. There are risks associated with the ecological (climatic uncertainty) and socioeconomic (socioeconomic uncertainty) contexts of the program. It also requires that we assess the impact of future management actions (technical uncertainty) on important attributes (Wyant *et al.*, 1995). Several formal failure/risk analyses are available for wildland development programs (Wyant *et al.*, 1995), but are not covered in this discussion. Climatic, technological, and socioeconomic uncertainties complicate the planning of wildland repair projects. Small projects applying well-established technologies in relatively predictable environments are less subject to problems caused by these uncertainties. However, their potential impact on project success increases dramatically as projects involve (1) more area; (2) more time; (3) untested technologies; and (4) greater reliance on socioeconomic systems. While planning cannot eliminate these uncertainties, their negative impact can be reduced with (1) more knowledge; (2) greater project flexibility; and (3) management that encourages innovation (Ffolliott *et al.*, 1994).

CLIMATIC UNCERTAINTY

While we cannot make the weather more predictable, we can reduce risks associated with climatic uncertainties by applying appropriate climatic information and technologies to reduce the adverse impact of climatic limitations. Climatic uncertainties require that we also consider infrequent, or extreme, events rather than simply planning for average conditions. Do not select species that are susceptible to temperature extremes expected within any 10 to 20 year period. It is also important to use the right kinds of climatic information rather than information that is easiest to obtain. For example, precipitation in arid and semiarid environments will be below average in 60–75% of the years. Thus, since plans requiring average precipitation are likely to fail more often than succeed, median precipitation is more useful for planning purposes. Climatic uncertainties can be addressed (1) by using many species, each with broad climatic adaptations; (2) with seedbed preparation and planting technologies that reduce the impact of climatic problems (drought, temperature extremes, wind); and (3) by designing the landscape to encourage positive microenvironmental interactions among species.

TECHNICAL UNCERTAINTY

Although planners can roughly estimate the anticipated benefits of a specific technology, uncertainties increase with new technologies (Ffolliott *et al.*, 1994). Unexpected results also occur when equipment or technologies are misapplied. Unanticipated, negative interactions among otherwise appropriate technologies will occur. Reducing our reliance on any single technology, effective training of the work force and proactive equipment maintenance also reduce uncertainties. Where little information exists about species adaptations, diverse species mixtures reduce the risk of complete failure. Adaptive management strategies, based on monitoring programs that provide early detection of problems, allow for modification or elimination of ineffective technologies. Large wildland repair projects should be implemented over many years, so that lessons learned from earlier work will improve subsequent efforts (Pastorok *et al.*, 1997). Applying adaptive management strategies to wildland repair programs requires flexible goals and designs, in addition to a long-term commitment to monitoring.

SOCIOECONOMIC UNCERTAINTIES

Socioeconomic uncertainties are usually greater than technical uncertainties. It is difficult to estimate current income, wealth, and education, but extrapolating these socioeconomic indicators into the future is even less reliable. This problem is greater in developing countries, where socioeconomic changes are more rapid. Where survival is a continuing concern, unsustainable landuse practices accelerate due to the pressures of low food-security and poverty (Kessler & Laban, 1994). Under these conditions, landuse decisions seldom consider the long-term implications.

PLANNING FOR RISK AND UNCERTAINTY

The uncertainties inherent in predicting future conditions, or the results of repair actions, should prevent narrow goals. It is also desirable to design heterogeneity into the physical, chemical, and biological components as a bet-hedging strategy (Pastorok *et al.*, 1997). Bet-hedging strategies emphasize the inclusion of spatial diversity and flexibility into the repair design as a means of establishing a system that is less limited by our misunderstandings and is more capable of adapting to future conditions (Pastorok *et al.*, 1997). Design features such as functional redundancy (see Chapter 5) reduce the risks associated with species establishment and the repair of damaged processes. Incorporating spatial variation into wildland landscapes has at least two important benefits (Pastorok *et al.*, 1997). First, different landscape components increase the probability that at least one will be successful and spread to adjacent, less-successful areas. These landscape components may differ in species, species ratios, age, seedbed preparation, planting techniques, or a variety of other attributes. Second, even if most of these subdesigns fail, this approach provides adaptive management information that will improve future repair efforts (Pastorok *et al.*, 1997).

Several strategies reduce the impact of uncertainties on repair programs (Ffolliott *et al.*, 1994). Obtaining additional knowledge before acting on unproven technologies reduces mistakes. Sensitivity analysis determines which factors have the most impact on project success. That allows a planning focus on the most critical factors. Planning methods that imagine alternative futures (and their consequences) reduce surprises and mistakes (Ffolliott *et al.*, 1994). Wildland repair programs that incorporate greater flexibility are less susceptible to unanticipated changes. Diversification by restricting planning to short intervals, monitoring the impacts, and using adaptive management will increase flexibility (Ffolliott *et al.*, 1994). Contingency planning helps formulate responses to new problems. While detailed, long-range plans are useful, uncertainty is most effectively addressed with innovative people operating with considerable autonomy (Ffolliott *et al.*, 1994).

Management intervention

A systematic approach to evaluating alternative repair strategies helps to consider the information developed to this point. The Society for Ecological Restoration (SER, 1994) suggested that restoration plans should contain, at a minimum, the following items:

1. A baseline ecological description of the ecosystem designated for restoration.
2. An evaluation of how the proposed restoration integrates with other parts of the regional landscape.
3. Explicit plans and schedules for all onsite preparation and installation activities.
4. Well-developed and explicit performance standards for evaluating the project.
5. Monitoring protocols for the performance standards.
6. Provision for the procurement of suitable plant materials and for supervision to guarantee proper planting.
7. Procedures for prompt post-installation maintenance and remediation.

After setting preliminary goals and operational treatment alternatives, we can begin to develop specific priorities, and final goals, and select from among various treatment options to develop the overall repair plan (Figure 1.5). Repair alternatives developed from the previous chapters contribute to the final plan. The suggested approach places initial emphasis on site stabilization and repairing damaged primary processes by initiating autogenic processes leading to the development of self-repairing ecosystems. Priority repair alternatives reduce resource loss while providing the desired goods and services. Human-caused degradation is assessed independently from normal cyclic instabilities (Wyant *et al.*, 1995). Although estimates of system-wide risk associated with both natural and human-caused disturbances are seldom very precise (Wyant *et al.*, 1995), they provide approximations suitable for planning. Effective monitoring and evaluation programs contribute feedback that reduces risk for future actions.

Monitoring and evaluation

Monitoring involves data collection, evaluation, and the analysis of that data to assess project success. Monitoring provides the information feedback that improves specific repair strategies (adaptive management). This continuing management change responds to measured results to improve future efforts (Ffolliott *et al.*, 1994; Wyant *et al.*, 1995). Thus, monitoring and evaluation are iterative processes that provide the feedback necessary for improvement. Ideally, the evaluation helps determine progress towards important goals and provides early warning signs of potential problems (Ffolliott *et al.*, 1994). Adaptive management does not suggest the original goals and methods were ill conceived (Wyant *et al.*, 1995). Rather, it reflects the complexity of wildland ecosystems and indicates the importance of flexibility to long-term projects. Monitoring and evaluation are especially important for large, long-term development projects. Many development projects consist of multiyear programs that depend on highly organized monitoring efforts for feedback to adaptive management (Wyant *et al.*, 1995).

Monitoring involves direct measurements, indirect measurements, imputed measurements, or some combination of the three (Ffolliott *et al.*, 1994). Direct measurements of important variables are most common. Predictive relationships derived from inputs and outputs produce indirect measurements. Imputed measurements derived from indices of both internal and external changes also contribute to monitoring data activities.

Information is expensive to collect, store, retrieve, and analyze. It is best to design monitoring programs early in the planning stage. Information users should be consulted to determine their requirements and the information necessary to make reasonably good decisions (Ffolliott *et al.*, 1994). Effective monitoring programs collect the necessary information in sufficient, but not excessive detail. Establishing priorities is essential to obtain the most important information and meet the needs of users in an efficient manner. Planning the monitoring program should include an analysis of the time and costs of obtaining the information, personnel training requirements, and report preparation (Ffolliott *et al.*, 1994).

It is seldom possible to monitor an exact duplicate ecosystem. The two most common types of reference information are historical data from the repair site and contemporary data from sites identified as a close match to the repair site (White & Walker, 1997). Numerous unmeasured factors complicate the interpretation of historical data. Finding reference sites that closely match the repair site is very difficult. Often, the only possible assessment is an onsite comparison of ecological and socioeconomic conditions before and after treatment. Where possible, it is preferable to locate similar sites that are either undamaged or similarly damaged and will not be repaired. The first provides a general target and the second a measure of change since the project started. Both provide information relating to the accomplishments of the repair program.

The response of any site to management, natural recovery processes, or repair actions is influenced by its history, management, and connections with the adjoining landscape (Pickett & Parker, 1994). Thus, the process and context emphasis (Figure 8.1) is a means of describing 'contingency', which suggests a whole range of possible results are possible. Extinctions, invasions by aggressive species, previous management, irreversible site degradation, and unique combinations of environmental conditions complicate the restoration of historical vegetation. Even if we know what the vegetation was at some point in the past, our goals should reflect a more dynamic view of ecosystem structure and function (White & Walker, 1997). Although setting narrowly defined conditions as goals may be unrealistic, reference ecosystems of similar landform/soil/biota/climate provide important guidance during a repair program, if our goals are flexible enough to allow for natural expression of a range of potential conditions.

Parameters relating to the conservation of limiting resources and the functioning of essential primary processes should receive priority. For example, if preliminary assessments suggested dysfunctional hydrologic processes, monitoring should emphasize parameters such as infiltration rate, runoff rates, sedimentation losses, water table changes, soil surface features, and plant cover. 'Vital ecosystem attributes' (Aronson *et al.*, 1993a; Aronson *et al.*, 1993b) and 'vital landscape attributes' (Aronson & Le Floc'h, 1996) provide a useful starting point for identifying parameters to be monitored (Hobbs & Norton, 1996). While ecological factors are included in all projects, many projects require a careful assessment of progress toward socioeconomic goals.

Numerous specific attributes have been mentioned as valuable assessments of wildland repair success (Tables 2.1, 2.2, 2.3, 2.4), yet few generalizations have emerged. For each attribute, repaired ecosystems are compared with reference ecosystems with similarity indices (Berger, 1991; Westman, 1991; Kondolf, 1995; Kondolf & Micheli, 1995). An alternative approach is based on the structural, compositional, and functional measurements originally proposed for the assessment of 'ecosystem health' (Costanza *et al.*, 1992). The approach assesses current conditions against the estimated range of natural variability of any relevant parameter (Hobbs & Norton, 1996). This assessment strategy could be expanded by developing ecological reference templates that define a limited range of functional and structural states (Allen, 1994). Then the current state of essential parameters for a site could be compared with the estimated range of natural variability for those parameters (Caraher & Knapp, 1995).

Literature cited

Abbott, I., C. A. Parker & I. D. Sills. (1979). Changes in the abundance of large soil animals and physical properties of soils following cultivation. *Australian Journal of Soil Research*, **17**, 345–353.

Aber, J. D. & J. M. Melillo. (1982). Nitrogen immobilization in decaying hardwood leaf litter as a function of initial nitrogen and lignin content. *Canadian Journal of Botany*, **60**, 2263–2269.

Aerts, R. & F. Berendse. (1988). The effect of increased nutrient availability on vegetation dynamics in wet heathlands. *Vegetatio*, **76**, 63–69.

Aerts, R., A. Huiszoon, J. H. A. Vanoostrum, C. A. D. M. Vandevijer & J. H. Willems. (1995). The potential for heathland restoration on formerly arable land at a site in Drenthe, The Netherlands. *Journal of Applied Ecology*, **32**, 827–835.

Aerts, R. & M. J. v. d. Peijl. (1993). A simple model to explain the dominance of low-productive perennials in nutrient-poor habitats. *OIKOS*, **66**, 144–147.

Agassi, M., I. Shainberg & J. Morin. (1981). Effect of electrolyte concentration and soil sodicity on infiltration rate and crust formation. *Soil Science Society of America Journal*, **45**, 848–851.

Ahmed, H. A. (1986). Some aspects of dry land afforestation in the Sudan with special reference to *Acacia tortilis* (Forsk.) Hayne, *Acacia senegal* Wild. and *Prosopis chilensis* (Molina) Stutz. *Forest Ecology and Management*, **16**, 209–221.

Alban, D. H. (1982). Effect of nutrient accumulation by aspen, spruce, and pine on soil properties. *Soil Science Society of America Journal*, **46**, 853–861.

Alexander, M. (1977). *Introduction to Soil Microbiology*, New York: John Wiley and Sons.

Allaby, M. (1994). *The Concise Oxford Dictionary of Ecology*, Oxford: Oxford University Press.

Allen, C. D. (1994). Ecological perspectives in linking ecology, GIS, and remote sensing to ecosystem management. In *Remote Sensing and GIS in Ecosystem Management*, ed. V. A. Sample, pp. 111–139. Washington, D.C.: Island Press.

Allen, E. (1989). The restoration of disturbed arid landscapes with special reference to mycorrhizal fungi. *Journal of Arid Environments*, 17, 279–286.

Allen, E. B. (1988a). *Introduction*. In *The Reconstruction of Disturbed Arid Lands. An Ecological Approach*, ed. E. B. Allen, pp. 1–4. Boulder, Colorado: Westview Press.

Allen, M. F. (1988b). Belowground structure: a key to reconstructing a productive arid ecosystem. In *The Reconstruction of Disturbed Arid Lands: An Ecological Approach*, ed. E. B. Allen, pp. 113–135. Boulder, Colorado: Westview Press.

Allen, M. F. & J. A. MacMahon. (1985). Impacts of disturbance on cold desert fungi: a comparative microscale dispersion patterns. *Pedobiologia*, 28, 215–224.

Allen, T. F. & T. W. Hoekstra. (1992). *Toward a Unified Ecology*, New York: Columbia University Press.

Anable, M. E., M. P. McClaran & G. P. Ruyle. (1992). Spread of introduced Lehman lovegrass (*Eragrostis lehmanniana* Nees.) in southern Arizona, USA. *Biological Conservation*, 61, 181–188.

Andelt, W. F., K. P. Burnham & J. A. Manning. (1991). Relative effectiveness of repellents for reducing mule deer damage. *Journal of Wildlife Management*, 55, 341–347.

Andersen, A. N. & G. P. Sparling. (1997). Ants as indicators of restoration success – relationship with soil microbial biomass in the Australian seasonal tropics. *Restoration Ecology*, 5, 109–114.

Anderson, D. C., K. T. Harper & R. C. Holmgren. (1982). Factors influencing development of cryptogamic soil crusts in Utah deserts. *Journal of Range Management*, 35, 180–185.

Anderson, R. (1981). Technology for reversing desertification. *Rangelands*, 3, 48–50.

Anderson, R. C. & K. J. Roberts. (1993). Mycorrhizae in prairie restoration: response of three little bluestem (*Schizachyrium scoparium*) populations to mycorrhizal inoculum from a single source. *Restoration Ecology*, 1, 83–87.

Andrews, J. H. & R. F. Harris. (1986). r- and K-selection and microbial ecology. *Advances in Microbial Ecology*, 9, 99–148.

Apfelbaum, S. I., B. J. Bader, F. Faessler & D. Mahler. (1997). Obtaining and processing seed. In *The Tallgrass Restoration Handbook: for Prairies, Savannas, and Woodlands*, ed. S. Packard & C. F. Mutel, pp. 99–134. Washington, D. C.: Island Press.

Archer, S. (1989). Have southern Texas savannas been converted to woodlands in recent history? *American Naturalist*, **134**, 545–561.

Archer, S. & D. A. Pyke. (1991). Plant-animal interactions affecting plant establishment and persistence on revegetated rangeland. *Journal of Range Management*, **44**, 558–565.

Archer, S. & F. E. Smeins. (1991). Ecosystem-level processes. *In Grazing Management: an ecological perspective*, ed. R. K. Heitschmidt & J. W. Stuth, pp. 109–139. Portland, Oregon: Timber Press.

Armbrust, D. V. & J. D. Bilbro. (1997). Relating plant canopy characteristics to soil transport capacity by wind. *Agronomy Journal*, **89**, 157–162.

Aronson, J., S. Dhillon & E. L. Floc'h. (1995). On the need to select an ecosystem of reference, however imperfect: a reply to Pickett and Parker. *Restoration Ecology*, **3**, 1–3.

Aronson, J., C. Floret, E. Le Floc'h, C. Ovalle & R. Pontanier. (1993a). Restoration and rehabilitation of degraded ecosystems in arid and semi-arid lands. I. A view from the south. *Restoration Ecology*, **1**, 8–17.

Aronson, J., C. Floret, E. Le Floc'h, C. Ovalle & R. Pontanier. (1993b). Restoration and rehabilitation of degraded ecosystems in arid and semi-arid lands. II. Case studies in southern Tunisia, Central Chile, and Northern Cameroon. *Restoration Ecology*, **1**, 168–187.

Aronson, J. & E. Le Floc'h. (1996). Vital landscape attributes – missing tools for restoration ecology. *Restoration Ecology*, **4**, 377–387.

Aronson, J., C. Ovalle & J. Avendano. (1993c). Ecological and economic rehabilitation of degraded 'Espinales' in the subhumid Mediterranean-climate region of central Chile. *Landscape and Urban Planning*, **24**, 15–21.

Aronson, J., C. Ovalle & J. Avendaño. (1992). Early growth rate and nitrogen fixation potential in forty-four legume species grown in an acid and a neutral soil from central Chile. *Forest Ecology and Management*, **47**, 225–244.

Ash, H. J., R. P. Gemmell & A. D. Bradshaw. (1994). The introduction of native plant species on industrial waste heaps: a test of immigration and other factors affecting primary succession. *Journal of Applied Ecology*, **31**, 74–84.

Ashby, W. C. (1997). Soil ripping and herbicides enhance tree and shrub restoration on stripmines. *Restoration Ecology*, **5**, 169–177.

Ashton, P. M. S., S. Gamage, I. A. U. N. Gunatilleke & C. V. S. Gunatilleke. (1997). Restoration of a Sri Lankan rainforest – using Caribbean pine *Pinus caribaea* as a nurse for establishing late-successional tree species. *Journal of Applied Ecology*, **34**, 915–925.

Austin, M. P. & O. Williams. (1988). Influence of climate and community composition on the population demography of pasture species in semi-arid Australia. *Vegetatio*, **77**, 43–49.

Bainbridge, D. A., M. Fidelibus & R. MacAller. (1995). Techniques for plant establishment in arid ecosystems. *Restoration and Management Notes*, 13, 190–197.

Bamforth, S. S. (1988). Interactions between Protozoa and other organisms. *Agriculture, Ecosystems and Environment*, 24, 225–234.

Banerjee, A. K. (1990). Revegetation technologies. In *Watershed Development in Asia: Strategies and Technologies*, ed. J. B. Doolette & W. B. Magrath, pp. 109–129. Washington, D.C.: The World Bank.

Barnett, J. P. (1991). *Production of shortleaf pine seedlings*. General Technical Report SO-90, Little Rock, Arkansas: U.S. Department of Agriculture, Forest Service, Southern Forest Experiment Station.

Barnett, J. P. & J. B. Baker. (1991). Regeneration methods. In *Forest Regeneration Manual*, ed. M. L. Duryea & P. M. Dougherty, pp. 35–50. Boston: Kluwer Academic Publishers.

Barro, S. C. & S. G. Conard. (1987). *Use of ryegrass seeding as an emergency revegetation measure in chaparral ecosystems*. General Technical Report PSW-102. Berkeley, California: U.S. Department of Agriculture, Forest Service, Pacific Southwest Forest and Range Experiment Station.

Barrow, C. J. (1991). *Land Degradation*, New York: Cambridge University Press.

Barrow, J. R. & K. M. Havstad. (1992). Recovery and germination of gelatin-encapsulated seeds feed to cattle. *Journal of Arid Environments*, 22, 395–399.

Barth, R. C. & J. O. Klemmedson. (1978). Shrub-induced spatial patterns of dry matter, nitrogen and organic carbon. *Soil Science Society of America Proceedings*, 42, 804–809.

Bartholomew, P. W., D. L. Easson & D. M. B. Chestnutt. (1981). A comparison of methods of establishing perennial and Italian ryegrasses. *Grass and Forage Science*, 36, 75–80.

Bartoszewica, A. & L. Ryszkowski. (1989). Influence of shelterbelts and meadows on the chemistry of ground water. In *Dynamics of Agricultural Landscapes*, ed. L. Ryszkowski, New York: Springer-Verlag.

Bell, M. J., B. J. Bridge, G. R. Harch & D. N. Orange. (1997). Physical rehabilitation of degraded krasnozems using ley pastures. *Australian Journal of Soil Research*, 35, 1093–1113.

Belnap, J. (1993). Recovery rates of cryptobiotic crusts: inoculant use and assessment methods. *Great Basin Naturalist*, 53, 89–95.

Belnap, J. & D. A. Gillette. (1997). Disturbance of biological soil crusts: impacts on potential wind erodibility of sandy desert soils in southeastern Utah. *Land Degradation and Development*, 8, 355–362.

Bement, R. E., R. D. Barmington, A. C. Everson, L. O. H. Jr. & E. E. Remenga. (1965). Seeding of abandoned croplands in the central Great Plains. *Journal of Range Management*, 18, 53–59.

Benjamin, R.W., D. Berkai,Y. Hofetz, M. A.Teifert &Y. Lavie. (1988). *Standing biomass of three species of fodder shrubs planted at five different densities, three years after planting.* Annual Report 6: 4–46. Beer-Sheva: Institute of Applied Research, Ben Gurion University of the Negev.

Bentham, H., J. A. Harris, P. Birch & K. C. Short. (1992). Habitat classification and soil restoration assessment using analysis of soil microbiological and physico-chemical characteristics. *Journal of Applied Ecology,* **29**, 711–718.

Berg, W. A. (1980). Nitrogen and phosphorus fertilization of mined lands. *Symposium on Adequate Reclamation of Mined Lands,* pp. 201–208. Billings, Montana: Soil Conservation Society of America.

Berg, W. A., J. W. Naney & S. J. Smith. (1991). Salinity, nitrate and water in rangeland and terraced wheatland above saline seeps. *Journal of Environmental Quality,* **20**, 8–11.

Berger, J. J. (1991). A generic framework for evaluating complex restoration and conservation projects. *Environmental Professional,* **13**, 254–262.

Berry, C. R. (1985). Subsoiling and sewage sludge aid loblolly pine establishment on adverse sites. *Reclamation and Revegetation Research,* **3**, 301–311.

Bertness, M. D. & R. Callaway. (1994). Positive interactions in communities. *TREE (Trends in Ecology and Evolution),* **9**, 191–193.

Bethlenfalvay, G. J. & S. Dakessian. (1984). Grazing effect on mycorrhizal colonization and floristic composition of the vegetation on a semiarid range in northern Nevada. *Journal of Range Management,* **37**, 312–316.

Bilbro, J. D. & D.W. Fryear. (1994). Wind erosion losses as related to plant silhoutte and soil cover. *Agronomy Journal,* **86**, 550–553.

Biondini, M., D. A. Klein & E. F. Redente. (1988). Carbon and nitrogen losses through root exudates by *Agropyron spicatum, A. smithii, and Bouteloua gracilis. Soil Biology and Biochemistry,* **20**, 477–482.

Biondini, M. & E. F. Redente. (1986). Interactive effect of stimulus and stress on plant community diversity in reclaimed lands. *Reclamation and Revegetation Research,* **4**, 211–222.

Bird, P. R., D. Bicknell, P. A. Bulman, S. J. A. Burke, J. F. Leys, J. N. Parker, F. J. v. d. Sommen & P.Volker. (1992).The role of shelter in Australia for protecting soils, plants and livestock. *Agroforestry Systems,* **20**, 59–86.

Bishop, S. C. & F. S. Chapin. (1989a). Establishment of *Salix alaxensis* on a gravel pad in arctic Alaska. *Journal of Applied Ecology,* **26**, 575–583.

Bishop, S. C. & F. S. Chapin. (1989b). Patterns of natural revegetation on abandoned gravel pads in arctic Alaska. *Journal of Applied Ecology,* **26**, 1073–1081.

BLM. (1973). *Determination of erosion condition class.* Form 7310–12. Washington, D. C.: U. S. Bureau of Land Management, Department of the Interior.

BLM. (1993). *Riparian area management: process for assessing proper functioning condition*. Technical Reference TR 1739–9 1993. Denver, Colorado: U. S. Department of Interior, Bureau of Land Management.

Bloomfield, H. E., J. F. Handley & A. D. Bradshaw. (1982). Nutrient deficiencies and the aftercare of reclaimed derelict land. *Journal of Applied Ecology*, 19, 151–158.

Bock, C. E., J. H. Bock, K. L. Jepsen & J. C. Ortega. (1986). Ecological effects of planting African lovegrasses in Arizona. *National Geographic Research*, 2, 456–463.

Boers, T. M. & J. Ben-Asher. (1982). A review of rainwater harvesting. *Agricultural Water Management*, 5, 145–158.

Borchers, S. & D. A. Perry. (1987). Early successional hardwoods as refugia for ectomycorrhizal fungi in clearcut Douglas fir forests of southwest Oregon. In *Mycorrhizae in the Next Decade: Practical Applications and Research Priorities*, ed. D. M. Sylvia, L. L. Hung & J. H. Graham, pp. 84. Gainsville, Florida: University of Florida.

Bosatta, E. & H. Staaf. (1982). The control of nitrogen turn-over in forest litter. *OIKOS*, 39, 143–151.

Bowman, R. A., J. D. Reeder & R. W. Lober. (1990). Changes in soil properties in a central plains rangeland after 3, 20, and 60 years of cultivation. *Soil Science*, 150, 851–857.

Bradshaw, A. D. (1983). The reconstruction of ecosystems. *Journal of Applied Ecology*, 20, 1–17.

Bradshaw, A. D. (1996). Underlying principles of restoration. *Canadian Journal of Fisheries and Aquatic Science*, 53, 3–9.

Bradshaw, A. D. (1997). What do we mean by restoration? In *Restoration Ecology and Sustainable Development*, ed. K. M. Urbanska, N. R. Webb & P. J. Edwards, pp. 8–14. Cambridge: Cambridge University Press.

Bradshaw, A. D. & M. J. Chadwick. (1980). *The Restoration of Land*, Oxford: Blackwell Scientific Publications.

Bradshaw, A. D., W. S. Dancer, J. F. Handley & J. C. Sheldon. (1975). The biology of land revegetation and the reclamation of the china clay wastes of Cornwall. In *The Ecology of Resource Degradation and Renewal*, ed. M. J. Chadwick & G. T. Goodman, pp. 363–384. Oxford: Blackwell Scientific Publications.

Brady, N. C. (1990). *The Nature and Properties of Soils*, New York: MacMillan Publishing Company.

Brakensiek, D. L. & W. J. Rawls. (1983). Agricultural management effects on soil water processes. Part II. Green and Ampt parameters for crusting soils. *Transactions of the American Society of Agricultural Engineers*, 26, 1753–1757.

Breedlow, P. A., P. V. Voris & L. E. Rogers. (1988). Theoretical perspective on ecosystem disturbance and recovery. In *Shrub-Steppe: Balance and Change in a Semi-Arid Terrestrial Ecosystem*, ed. W. H. Rickard, L. E. Rogers, B. E. Vaughan & S. F. Liebetrau, pp. 257–269. New York: Elsevier.

Brissette, J. C., J. P. Barnett & T. D. Landis. (1991). Container seedlings. In *Forest Regeneration Manual*, ed. M. L. Duryea & P. M. Dougherty, pp. 117–141. Boston: Kluwer Academic Publishers.

Brooks, K., P. F. Ffolliott, H. M. Gregersen & J. L. Thames. (1991). *Hydrology and the Management of Watersheds*, Ames, Iowa: Iowa State University Press.

Brown, J. H. & E. J. Heske. (1990). Control of a desert-grassland transition by a keystone rodent guild. *Science*, **250**, 1705–1707.

Brown, J. R. & S. Archer. (1987). Woody plant dispersal and gap formation in a North American subtropical savanna woodland: the role of domestic herbivores. *Vegetatio*, **73**, 73–80.

Brown, R. W., R. S. Johnston & D. A. Johnson. (1978). Rehabilitation of alpine tundra disturbances. *Journal of Soil and Water Conservation*, **33**, 154–160.

Brown, S. & A. E. Lugo. (1994). Rehabilitation of tropical lands: A key to sustaining development. *Restoration Ecology*, **2**, 97–111.

Bruns, D. (1988). Restoration and management of ecosystems for nature conservation in West Germany. In *Rehabilitating Damaged Ecosystems*, ed. J. Cairns, Jr., pp. 183–186. Boca Raton, Florida: CRC Press.

Bryan, R. B. (1979). The influence of slope angle on soil entrainment by sheetwash and rainsplash. *Earth Surface Processes and Landforms*, **4**, 43–58.

Bryant, J. P. & F. S. Chapin. (1986). Browsing-woody plant interactions during boreal forest plant succession. In *Forest Ecosystems in the Alaskan Taiga*, ed. K. Van Cleve, F. S. Chapin, P. W. Viereck, L. A. Dryness & C. R. Dryness, pp. 213–225. New York: Springer-Verlag.

Bryant, J. P., P. B. Reichardt & T. P. Clausen. (1992). Chemically mediated interactions between woody plants and browsing animals. *Journal of Range Management*, **45**, 18–24.

Bullock, J. M. & N. R. Webb. (1995). A landscape approach to heathland restoration. In *Restoration Ecology in Europe*, ed. K. M. Urbanska & K. Grodzinska, pp. 71–91. Zürich: Geobotanical Institute SFIT Zürich.

Burel, F., J. Baudry & J. Lefeuvre. (1993). Landscape structure and the control of water runoff. In *Landscape Ecology and Agroecosystems*, ed. R. G. H. Bunce, L. Ryszkowski & M. G. Paoletti, pp. 41–47. Boca Raton, Florida: Lewis Publishers.

Burke, I. C., W. K. Laurenroth & D. P. Coffin. (1995). Soil organic matter recovery in semiarid grasslands: implications for the conservation reserve program. *Ecological Applications*, **5**, 793–801.

Burrows, W. H. (1991). Sustaining productive pastures in the tropics. 11. An ecological perspective. *Tropical Grasslands*, **25**, 153–158.

Burton, G. W. & J. S. Andrew. (1948). Recovery and viability of seed of certain southern grasses and lespedeza passed through bovine digestive tract. *Journal of Agricultural Research*, **76**, 95–103.

Burton, G. W. & W. W. Hanna. (1985). Bermudagrass. In Forages: *The Science of Grassland Agriculture*, ed. M. E. Heath, R. F. Barnes & D. S. Metcalfe, pp. 247–254. Ames, Iowa: Iowa State University Press.

Cairns, J., Jr. (1988). Restoration ecology: the new frontier. In *Rehabilitating Damaged Ecosystems*, ed. J. Cairns, Jr., pp. 1–12. Boca Raton, Florida: CRC Press, Inc.

Cairns, J., Jr. (1989). Restoring damaged ecosystems: is predisturbance condition a viable option? *Environmental Professional*, **11**, 152–159.

Cairns, J., Jr. (1991). The status of the theoretical and applied science of restoration ecology. *Environmental Professional*, **13**, 1–9.

Campbell, D. L. & J. Evans. (1975). *"Vexar" seedling protectors to reduce wildlife damage to Douglas-fir*. Wildlife Leaflet 508. Washington, DC: United States Department of the Interior, Fish and Wildlife Service.

Caraher, D. & W. H. Knapp. (1995). *Assessing ecosystem health in the Blue Mountains*. General Technical Report SE-88. Hendersonville, North Carolina: U.S. Department of Agriculture, Forest Service, Southeast Forest Experiment Station.

Carr, W. W. & T. M. Ballard. (1980). Hydroseeding forest roadsides in British Columbia for erosion control. *Journal of Soil and Water Conservation*, **35**, 33–35.

Carson, W. P. & G. W. Barrett. (1988). Succession in old-field communities: effects of contrasting types of nutrient enrichment. *Ecology*, **69**, 984–994.

Casenave, A. & C. Valentin. (1992). A runoff capability classification system based on surface features criteria in semi-arid areas of West Africa. *Journal of Hydrology*, **130**, 231–249.

Chambers, J. C. (1989). *Native species establishment on an oil drill pad site the Unitah Mountains, Utah: effects of introduced grass density and fertilizer*. Research Paper INT-402. Ogden, Utah: U.S. Department of Agriculture, Forest Service, Intermountain Research Station.

Chambers, J. C. (1995). Relationships between seed fates and seedling establishment in an alpine ecosystem. *Ecology*, **76**, 2124–2133.

Chapin, F. S., III. (1980). The mineral nutrition of wild plants. *Annual Review of Ecology and Systematics*, **11**, 233–260.

Chapin, F. S., III, P. M. Vitousek & K. V. Cleve. (1986). The nature of nutrient limitation in plant communities. *The American Naturalist*, **127**, 48–58.

Chapin, F. S., III, B. H. Walker, R. J. Hobbs, D. U. Hooper, J. H. Lawton, O. E. Sala & D. Tilman. (1997). Biotic control over the functioning of ecosystems. *Science*, **277**, 500–504.

Chase, R. & E. Boudouresque. (1987). Methods to stimulate plant regrowth on bare Sahelian forest soils in the region of Niamey, Niger. *Agriculture, Ecosystems and Environment*, **18**, 211–221.

Chen, Y., J. Tarchitzky, J. Bower, J. Morin & A. Banin. (1980). Scanning electron microscope observations on soil crusts and their formation. *Soil Science*, **130**, 49–55.

Chepil, W. S. (1955). Factors that influence clod structure and erodibility of soil by wind. V. Organic matter at various stages of decomposition. *Soil Science*, **80**, 413–421.

Chepil, W. S. & N. P. Woodruff. (1963). The physics of wind erosion and its control. *Advances in Agronomy*, **15**, 211–302.

Chiarello, N., J. C. Hichman & H. A. Mooney. (1982). Endomycorrhizal role for interspecific transfer of phosphorus in a community of annual plants. *Science*, **217**, 941–943.

Choi, Y. D. & M. K. Wali. (1995). The role of *Panicum virgatum* (switchgrass) in the revegetation of iron-mine tailings in northern New York. *Restoration Ecology*, **3**, 123–132.

Christy, E. J. & R. N. Mack. (1984). Variation in demography of juvenile *Tsuga hererophylla* across the substratum mosaic. *Journal of Ecology*, **72**, 75–91.

Clarke, C. T. (1997). Role of soils in determining sites for lowland heathland reconstruction in England. *Restoration Ecology*, **5**, 256–264.

Clary, W. P. (1989). *Revegetation by land imprinter and rangeland drill*. Research Paper INT-397. Ogden, Utah: U.S. Department of Agriculture, Forest Service, Intermountain Research Station.

Clements, F. E. (1916). *Plant Succession: an Analysis of the Development of Vegetation*. 242. Washington D.C.: Carnegie Institute.

Clements, F. E. (1936). Nature and the structure of the climax. *Journal of Ecology*, **24**, 252–284.

Clewell, A. F. & R. Lea. (1990). Creation and restoration of forested wetland vegetation in the southeastern United States. In *Wetland Creation and Restoration*, ed. J. A. Kusler & M. E. Kentula, pp. 195–231. Washington, D.C.: Island Press.

Coley, P. D. (1988). Effects of plant growth rate and leaf lifetime on the amount and type of anti-herbivore defense. *Oecologia*, **74**, 531–536.

Coley, P. D., J. P. Bryant & F. S. Chapin. (1985). Resource availability and plant antiherbivore defense. *Science*, **230**, 895–899.

Connell, J. H. & R. O. Slatyer. (1977). Mechanisms of succession in natural communities and their role in community stability and organization. *American Naturalist*, **111**, 1119–1144.

Coppin, N. & R. Stiles. (1995). Ecological principles for vegetation establishment and maintenance. In *Slope Stabilization and Runoff Control: A Bioengineering Approach*, ed. R. P. C. Morgan & R. J. Rickson, pp. 59–93. New York: E & FN Spon.

Costanza, R. (1992). Toward and operational definition of ecosystem health. In *Ecosystem Health: New Goals for Environmental Management*, ed. R. Costanza, B. G. Norton & B. D. Haskell, pp. 239–256. Washington, D.C.: Island Press.

Costanza, R., B. G. Norton & B. D. Haskell (eds) (1992). *Ecosystem Health: New Goals for Environmental Management*, Washington, D.C.: Island Press.

Cottam, G. (1987). Community dynamics on an artificial prairie. In *Restoration Ecology: a Synthetic Approach to Ecological Research*, ed. W. R. Jordan, M. E. Gilpin & J. D. Aber, pp. 257–270. Cambridge: Cambridge University Press.

Cotts, N. R., E. F. Redente & R. Schiller. (1991). Restoration methods for abandoned roads at lower elevations in Grand Teton National Park, Wyoming. *Arid Soil Research and Rehabilitation*, 5, 235–249.

Coughlin, K. J., W. E. Fox & J. D. Hughes. (1973). Aggregation in swelling clay soils. *Australian Soil Research*, 11, 133–141.

Cox, J. R. & M. H. Martin. (1984). Effects of planting depth and soil texture on the emergence of four lovegrasses. *Journal of Range Management*, 37, 204–205.

Cox, J. R., J. M. Parker & J. L. Stroelein. (1984). Soil properties in creosote bush communities and their relative effects on the growth of seeded range grasses. *Soil Science Society of America Journal*, 48, 1442–1445.

Crawley, M. J. (1983). *Herbivory: the Dynamics of Animal–Plant Interactions*, Berkeley, California: University of California Press.

Crocker, R. L. & J. Major. (1955). Soil development in relation to vegetation and surface age at Glacier Bay, Alaska. *Journal of Ecology*, 43, 427–448.

Cronk, Q. C. B. & J. L. Fuller. (1995). Plant Invaders: *The Threat to Natural Ecosystems*, New York: Chapman and Hall.

Cubbage, F. W., J. E. Gunter & J. T. Olson. (1991). Reforestation economics, law, and taxation. In *Forest Regeneration Manual*, ed. M. L. Duryea & P. M. Dougherty, pp. 9–31. The Hague, Netherlands: Kluwer Academic Publishers.

D'Antonio, C. M. & P. M. Vitousek. (1992). Biological invasions by exotic grasses, the grass/fire cycle, and global change. *Annual Review of Ecology and Systematics*, 23, 63–87.

Daley, H. E. (1991). *Steady-State Economics*, Washington, D.C.: Island Press.

Dalrymple, J. B., R. J. Blong & A. J. Conacher. (1968). A hypothetical nine-unit land surface model. *Zeitschrift für Geomorphologie*, 12, 60–76.

Dancer, W. S., J. F. Handley & A. D. Bradshaw. (1977). Nitrogen accumulation in kaolin wastes in Cornwell. II. Forage legumes. *Plant and Soil,* 48, 303–314.

Danin, A. (1991). Plant adaptations in desert dunes. *Journal of Arid Environments,* 21, 193–212.

Davenport, D. W., D. D. Breshears, B. P. Wilcox & C. G. Allen. (1998). Viewpoint: sustainability of piñion-juniper ecosystems – a unifying perspective of soil erosion thresholds. *Journal of Range Management,* 51, 231–240.

Davidson, D. W. (1993). The effects of herbivory and granivory on terrestrial plant succession. *OIKOS,* 68, 23–35.

Davidson, D. W., R. S. Inouye & J. H. Brown. (1984). Granivory in a desert ecosystem: experimental evidence for indirect facilitation of ants by rodents. *Ecology,* 65, 1780–1786.

Davidson, D. W. & D. A. Samson. (1985). Granivory in the Chihuahuan Desert: interactions within and between trophic levels. *Ecology,* 66, 486–502.

Davies, R., A. Younger & R. Chapman. (1992). Water availability in a restored soil. *Soil Use and Management,* 8, 67–73.

Dawson, J. O. (1986). Actinorhizal plants: their use in forestry and agriculture. *Outlook on Agriculture,* 15, 202–208.

De Vries, J. & T. L. Chow. (1978). Hydraulic behavior of a forested mountain soil in coastal British Columbia. *Water Resources Research,* 14, 933–935.

DeAngelis, D. L., W. M. Post & C. C. Travis. (1986). *Positive Feedback in Natural Systems,* Berlin: Springer-Verlag.

DeAngelis, D. L. & J. C. Waterhouse. (1987). Equilibrium and nonequilibrium concepts in ecological models. *Ecological Monographs,* 57, 1–21.

Debussche, M. & P. Isenmann. (1994). Bird-dispersed seed rain and seedling establishment in patchy Mediterranean vegetation. OIKOS, 69, 414–426.

Decker, A. M. & T. H. Taylor. (1985). Establishment of new seedings and renovation of old sods. In *Forages: The Science of Grassland Agriculture,* ed. M. E. Heath, R. F. Barnes & D. S. Metcalfe, pp. 288–297. Ames, Iowa: Iowa State University Press.

DeLeo, G. A. & S. Levin. (1997). The multifaceted aspects of ecosystem integrity. *Conservation Ecology,* [online] 1, 3.

DePuit, E. J. (1988a). Productivity of reclaimed lands – rangeland. In *Reclamation of Surface-Mined Lands,* ed. L. R. Hossner, pp. 93–129. Boca Raton, Florida: CRC Press, Inc.

DePuit, E. J. (1988b). Productivity of reclaimed lands – rangelands. In *Reclamation of Surface-Mined Lands,* ed. L. R. Hossner, pp. 93–130. Boca Raton, Florida: CRC Press.

Diamond, J. A. (1975a). Assembly of species communities. In *Ecology and Evolution of Communities*, ed. M. L. Cody & J. A. Diamond, pp. 342–444. Cambridge, MA: Harvard University Press.

Diamond, J. M. (1975b). The island dilemma: lessons of modern biogeographic studies for the design of nature reserves. *Biological Conservation*, 7, 129–146.

Diboll, N. (1997). Designing seed mixes. In *The Tallgrass Restoration Handbook: for Prairies, Savannas, and Woodlands*, ed. S. Packard & C. F. Mutel, pp. 135–149. Washington, D.C.: Island Press.

Dickerson, J. D., N. P. Woodruff & E. E. Banbury. (1976). Techniques for improving tree survival and growth in semiarid areas. *Journal of Soil and Water Conservation*, 31, 63–66.

Dixon, R. M. (1990). Land imprinting for dryland revegetation and restoration. In *Environmental Restoration: Science and Strategies for Restoring the Earth*, ed. J. J. Berger, pp. 14–22. Washington, D.C.: Island Press.

Dixon, R. M. & A. E. Peterson. (1971). Water infiltration control: a channel system concept. *Soil Science Society of America Proceedings*, 35, 968–973.

Dobson, A. P., A. D. Bradshaw & A. J. M. Baker. (1997). Hopes for the future – restoration ecology and conservation biology. *Science*, 277, 515–522.

Doescher, P. S., R. F. Miller & A. H. Winward. (1984). Soil chemical patterns under eastern Oregon plant communities dominated by big sagebrush. *Soil Science of America Journal*, 48, 659–663.

Dormaar, J. F., M. A. Nash, W. D. Williams & D. S. Chanasyk. (1995). Effect of native prairie, crested wheatgrass (*Agropyron cristatum* (L.) Gaertn.) and Russian wildrye (*Elymus junceus* Fisch.) on soil chemical properties. *Journal of Range Management*, 48, 258–263.

Downs, J. L., W. H. Rickard & L. L. Caldwell. (1993). Restoration of big sagebrush habitat in Southeastern Washington. *Wildland Shrub and Arid Land Restoration Symposium*, INT-GTR-315, ed. B. Roundy, E. D. McArthur, J. S. Haley & D. K. Mann, pp. 74–77. Las Vegas, Nevada: U.S. Department of Agriculture, Forest Service, Intermountain Research Station.

Drees, L. R., A. Manu & L. P. Wilding. (1993). Characteristics of eolian dusts in Niger, West Africa. *Geoderma*, 59, 213–233.

Duffy, P. D. & D. C. McClurkin. (1967). Stabilizing gully banks with excelsior mulch and loblolly pine. *Journal of Soil and Water Conservation*, 22, 70–71.

Eck, H. V., R. F. Dudley, R. H. Ford & J. C. W. Gantt. (1968). Sand dune stabilization along streams in the southern Great Plains. *Journal of Soil and Water Conservation*, 23, 131–134.

Edwards, W. M. (1991). Soil structure: processes and management. In *Soil Management for Sustainability*, ed. R. Lal & F. J. Pierce, pp. 7–14. Ankeny, Iowa: Soil and Water Conservation Society.

Egler, F. E. (1954). Vegetation science concepts. I. Initial floristic composition
– a factor in old-field vegetation development. *Vegetatio*, 4, 412–417.

Eissenstat, D. M. & E. I. Newman. (1990). Seedling establishment near large
plants: effects of vesicular-arbuscular mycorrhizaes on the intensity of
plant competition. Functional Ecology, 4, 95–99.

El Asswad, R. L., A. O. Said & M. T. Mornag. (1992). Effect of olive oil cake
on water holding capacity of sandy soils in Libya. *Journal of Arid
Environments*, 24, 409–413.

Eldridge, D. J. (1993a). Cryptogam cover and soil surface condition: effects on
hydrology on a semiarid woodland soil. *Arid Research and Rehabilitation*,
7, 203–217.

Eldridge, D. J. (1993b). Cryptogams, vascular plants, and soil hydrological
relations: some preliminary findings from the semiarid woodlands of
eastern Australia. *Great Basin Naturalist*, 53, 48–58.

Eldridge, D. J., M. E. Tozer & S. Slangen. (1997). Soil hydrology is independent
of microphytic crust cover – further evidence from a wooded semiarid
Australian rangeland. *Arid Soil Research & Rehabilitation*, 11, 113–126.

Eliason, S. A. & E. B. Allen. (1997). Exotic grass competition in suppressing
native shrubland reestablishment. *Restoration Ecology*, 5, 245–255.

Elkins, N. Z., Y. Steinberger & W. G. Whitford. (1982). The role of microarthropods and nematodes in decomposition in a semi-arid ecosystem.
Oecologia, 55, 303–310.

Elliot, E. T., R. J. Anderson, D. C. Coleman & C. V. Cole. (1980). Habitable
pore space and microbial trophic interactions. *OIKOS*, 35, 327–335.

Ellis, R. H., T. D. Hong & E. H. Roberts. (1989). A comparison of the low-
moisture content limit to the logarithmic relation between seed moisture
and longevity in twelve species. *Annals of Botany*, 63, 601–611.

Ellis, R. H., T. D. Hong & E. H. Roberts. (1990). An intermediate category of
seed storage behavior? I. Coffee. *Journal of Experimental Botany*, 41,
1167–1174.

Emmett, W. W. (1978). Overland flow. In *Hillslope Hydrology*, ed. M. J. Kirby,
pp. 145–176. New York: John Wiley and Sons.

Engman, E. T. (1986). Roughness coefficients for routing surface runoff.
*Journal of Irrigation and Drainage Division, American Society of Civil
Engineering*, 112, 39–53.

Evans, J. M. (1991). *Propagation of riparian species in southern California.*
General Technical Report RM-211. Park City, Utah: U.S. Department of
Agriculture, Forest Service, Rocky Mountain Forest and Range
Experiment Station.

Evans, R. A., H. R. Holbo, J. R. E. Eckert & J. A. Young. (1970). Functional
environment of downy brome communities in relation to weed control
and revegetation. *Weed Science*, 18, 154–162.

Evans, R. A. & J. A. Young. (1975). Enhancing germination of dormant seeds of downy brome. *Weed Science*, **23**, 354–357.

Evans, R. A. & J. A. Young. (1978). Effectiveness of rehabilitation practices following wildfire in a degraded big sagebrush-downy brome community. *Journal of Range Management*, **31**, 185–188.

Evans, R. A. & J. A. Young. (1984). Microsite requirements for downy brome (Bromus tectorum) infestation and control on sagebrush rangelands. *Weed Science*, **32**, 13–17.

Evans, R. D. & J. R. Ehleringer. (1993). A break in the nitrogen cycle in aridlands? Evidence from d15N of soils. *Oecologia*, **94**, 314–317.

Everett, R. L., R. O. Meewig & R. Stevens. (1978). Deer mouse preference for seed of commonly planted species, indigenous weed seed, and sacrifice foods. *Journal of Range Management*, **31**, 70–73.

Ewel, J. J. (1986). Designing agricultural ecosystems for the humid tropics. *Annual Review of Ecology and Systematics*, **17**, 245–271.

Ewel, J. J. (1997). Ecosystem processes and the new conservation theory. In *The Ecological Basis of Conservation: Heterogeneity, Ecosystems, and Biodiversity*, ed. S. T. A. Pickett, R. S. Ostfeld, M. Shachek & G. E. Likens, pp. 252–261. New York: Chapman & Hall.

Ewel, J. J., M. J. Mazzarino & C. W. Berish. (1991). Tropical soil fertility changes under monocultures and successional communities of different structure. *Ecological Applications*, **1**, 289–302.

Facelli, J. & S. T. A. Pickett. (1991). Indirect effects of litter on woody seedlings subject to herb competition. *OIKOS*, **62**, 129–138.

FAO. (1989). *Arid zone forestry: a guide for field technicians*. FAO Conservation Guide 20. Rome, Italy: Food and Agriculture Organization of the United Nations.

Farrell, J. (1990). The influence of trees in selected agroecosystems in Mexico. In *Agroecology: Researching the Ecological Basis for Sustainable Agriculture*, ed. S. R. Gliessman, pp. 167–183. New York: Springer-Verlag.

Felker, P., C. Wiesman & D. Smith. (1988). Comparison of seedling containers on growth and survival of *Prosopis alba* and *Leucaena leucocephala* in semi-arid conditions. *Forest Ecology and Management*, **24**, 177–182.

Ffolliott, P. F., K. N. Brooks, H. M. Gregersen & A. L. Lundgren. (1994). *Dryland Forestry: Planning and Management*, New York: John Wiley & Sons.

Fimbel, R. A. & C. C. Fimbel. (1996). The role of exotic conifer plantations in rehabilitating degraded tropical forest lands: a case study from the Kibale Forest in Uganda. *Forest Ecology and Management*, **81**, 215–226.

Finn, J. T. (1976). Measures of ecosystem structure and function derived from analysis of flows. *Journal of Theoretical Biology*, **56**, 363–380.

Fisher, J. T., G. A. Fancher & E. F. Aldon. (1990). Factors affecting establishment of one-seeded juniper (*Juniperus monosperma*) on surface-mined lands in New Mexico. *Canadian Journal of Forest Research*, 20, 880–886.

Flanagan, N. E., W. J. Mitsch & K. Beach. (1994). Predicting metal retention in a constructed mine drainage wetland. *Ecological Engineering*, 3, 135–159.

Flanagan, P. W. & K. V. Cleve. (1983). Nutrient cycling in relation to decomposition and organic matter quality in taiga ecosystems. *Canadian Journal of Forest Research*, 13, 95–817.

Fleming, L. V. (1983). Succession of mycorrhizal fungi on birch: infection of seedlings planted around mature trees. *Plant and Soil*, 71, 263–267.

Fleming, L. V. (1984). Effects of soil trenching and coring on the formation of ectomycorrhizas on birch seedlings grown around mature trees. *New Phytologist*, 98, 143–153.

Floret, C., E. L. Floc'h & R. Pontanier. (1990). Principles of zone identification and of interventions to stabilize sands in arid mediterranean regions. *Arid Soil Research and Rehabilitation*, 4, 33–41.

Forman, R. T. T. & M. Godron. (1986). *Landscape Ecology*, New York: John Wiley and Sons, Inc.

Fowler, D. K. & J. B. Maddox. (1974). Habitat improvement along reservoir inundation zones by barge hydroseeding. *Journal of Soil and Water Conservation*, 29, 263–265.

Fowler, N. L. (1986). Microsite requirements for germination and establishment of three grass species. *American Midland Naturalist*, 115, 131–145.

Fox, B. J. & J. H. Brown. (1993). Assembly rules for functional groups in North American desert rodent communities. *OIKOS*, 67, 358–370.

Fox, D. & R. B. Bryan. (1992). Influence of a polyacrylamide soil conditioner on runoff generation and soil erosion: field tests in Baringo District, Kenya. *Soil Technology*, 5, 101–119.

Franco, A. A. & S. M. Defaria. (1997). The contribution of N_2–fixing tree legumes to land reclamation and sustainability in the tropics. *Soil Biology & Biochemistry*, 29, 897–903.

Frankel, O. H. (1974). Genetic conservation: our evolutionary responsibility. *Genetics*, 78, 53–65.

Fresquez, P. R., E. F. Aldon & W. C. Lindermann. (1987). Enzyme activities in reclaimed coal mine spoils and soils. *Landscape and Urban Planning*, 14, 359–364.

Friedel, M. H. (1991). Range condition assessment and the concept of thresholds: a viewpoint. *Journal of Range Management*, 44, 422–426.

Frost, T. M., S. R. Carpenter, A. R. Ives & T. K. Kratz. (1995). Species compensation and complementarity in ecosystem functioning. In *Linking Species and Ecosystems*, ed. G. G. Jones & J. H. Lawton, pp. 224–239. London: Chapman and Hall.

Garcia-Moya, E. & C. M. McKell. (1970). Contribution of shrubs to the nitrogen economy of a desert wash plant community. *Ecology*, 51, 81–88.

Gardner, C. J. (1993). The colonization of a tropical grassland by *Stylosanthes* from seed transported in cattle feces. *Australian Journal of Agricultural Research*, 44, 299–315.

Garner, W. & Y. Steinberger. (1989). A proposed mechanism for the formation of 'fertile islands' in the desert ecosystem. *Journal of Arid Environments*, 16, 257–262.

Geber, M. & T. E. Dawson. (1993). Evolutionary responses of plants to global change. In *Biotic Interactions and Global Change*, ed. P. M. Kareiva, J. G. Kingsolver & R. B. Huey, pp. 179–197. Sunderland, Massachusetts: Sinaur Associates.

George, M. R., J. R. Brown & W. J. Clawson. (1992). Application of non-equilibrium ecology to management of Mediterranean grassland. *Journal of Range Management*, 45, 436–440.

Gillette, D. A., J. Adams, D. Smith & R. Kihl. (1980). Threshold velocities for input of soil particles into the air by desert soils. *Journal of Geophysical Research*, 85C, 5621–5630.

Gillette, D. A. & J. P. Dobrowolski. (1993). Soil crust formation by dust deposition at Shaartuz, Tadzhik, S.S.R. *Atmospheric Environment*, 27A, 2519–2525.

Girard, S., A. Clement, H. Cochard, B. Bouletgercourt & J. M. Guehl. (1997). Effects of desiccation on post-planting stress in bare-root Corsican pine seedlings. *Tree Physiology*, 17, 429–435.

Gleason, H. A. (1939). The individualistic concept of the plant association. *American Midland Naturalist*, 21, 92–108.

Goebel, C. J. & G. Berry. (1976). Selectivity of range grass seed by local birds. *Journal of Range Management*, 29, 393–395.

Good, L. G. & D. E. Smika. (1978). Chemical fallow for soil and water conservation in the Great Plains. *Journal of Soil and Water Conservation*, 33, 89–90.

Gosling, P. G. (1991). Beechnut storage: a review and practical interpretation of the scientific literature. *Forestry*, 64, 51–59.

GPAC. (1966). *A stand establishment survey of grass plantings in the Great Plains*. Great Plains Agricultural Council Report 23. Lincoln, Nebraska: Nebraska Agriculture Experiment Station.

Grainger, A. (1992). Characterization and assessment of desertification processes. In *Desertified Grasslands. Their Biology and Management*, ed. G. P. Chapman, pp. 17–33. New York: Academic Press.

Grant, P. F. & W. G. Nickling. (1998). Direct field measurement of wind drag on vegetation for application to windbreak design and modeling. *Land Degradation and Development*, 9, 57–66.

Greene, D. F. & E. A. Johnson. (1996). Wind dispersal of seeds from a forest into a clearing. *Ecology*, 77, 595–609.

Greenwood, E. A. N. (1988). The hydraulic role of vegetation in the development and reclamation of dryland salinity. In The *Reconstruction of Disturbed Arid Lands: An Ecological Approach*, ed. E. B. Allen, Boulder, Colorado: Westview Press.

Grime, J. P. (1977). Evidence for the expression of three primary strategies in plants and its relevance to ecological and evolutionary theory. *American Naturalist*, 111, 1169–1174.

Grime, J. P. (1979). *Plant Strategies and Vegetation Processes*, New York: John Wiley and Sons.

Grime, J. P. (1986). Manipulation of plant species and communities. In *Ecology and Design in Landscape*, ed. A. D. Bradshaw, D. A. Goode & E. H. P. Thorp, London: Blackwell Scientific Publications.

Grime, J. P. (1987). Mechanisms promoting floristic diversity in calcareous grasslands. *Proceedings of a Joint British Ecological Society/Nature Conservancy Council Symposium*, ed. S. H. Hiller, D. W. H. Walton & D. A. Wells, pp. 51–56. Sheffield, England: Bluntisham Books.

Grime, J. P. (1989). The stress debate: symptom of impending synthesis? *Biological Journal of the Linnean Society*, 37, 3–17.

Grime, J. P. & R. Hunt. (1975). Relative growth-rate: its range and adaptive significance in a local flora. *Journal of Ecology*, 63, 393–422.

Grossman, J. (1990). Mulch better. *Agrichemical Age*, 34, 4–5, 16–17.

Guariguata, M. R. & J. M. Dupuy. (1997). Forest regeneration in abandoned logging roads in lowland Costa Rica. *Biotropica*, 29, 15–28.

Guariguata, M. R., R. Rheingans & F. Montagnini. (1995). Early woody invasions under tree plantations in Costa Rica: implications for forest restoration. *Restoration Ecology*, 3, 252–260.

Guerrant, E. O., Jr. (1996). Designing populations: demographic, genetic, and horticultural dimensions. In *Restoring Diversity: Strategies for Reintroduction of Endangered Plants*, ed. D. A. Falk, C. I. Miller & M. Olwell, pp. 171–207. Washington, D.C.: Island Press.

Haila, Y., D. A. Saunders & R. J. Hobbs. (1993). What do we presently understand about ecosystem fragmentation? In *The Reconstruction of Fragmented Ecosystems*, ed. D. A. Saunders, R. J. Hobbs & P. R. Erlich, pp. 45–55. Chipping Norton: Surrey Beatty & Sons.

Hall, A. E., G. H. Cannell & H. W. Lawton. (1979). *Agriculture in Semi-arid Environments*, Berlin: Springer-Verlag.

Hansen, A. J., P. G. Risser & F. d. Castri. (1992). Epilogue: biodiversity and ecological flows across ecotones. In *Landscape Boundaries: Consequences for Biotic Diversity and Ecological Flows*, ed. A. J. Hansen & F. d. Castri, pp. 423–438. New York: Springer-Verlag.

Hardt, R. A. & R. T. T. Forman. (1989). Boundary effects on woody colonization of reclaimed surface mines. *Ecology*, 70, 1252–1260.

Harmer, R. & G. Kerr. (1995). Creating woodlands: to plant trees or not? In *The Ecology of Woodland Creation*, ed. R. Ferris-Kaan, pp. 113–128. New York: John Wiley & Sons.

Harper, J. L. (1977). *Population Biology of Plants*, New York: Academic Press.

Harper, J. L., M. Jones & N. R. S. Hamilton. (1992). The evolution of roots and the problems analysing their behavior. In *Plant Root Growth: An Ecological Perspective*, ed. D. Atkinson, pp. 3–22. Oxford: British Ecological Society.

Harper, J. L., J. T. Williams & G. R. Sagar. (1965). The behaviour of seeds in the soil: I. The heterogeneity of soil surfaces and its role in determining the establishment of plants from seed. *Journal of Ecology*, 53, 273–286.

Harrington, J. F. (1972). Seed storage and longevity. In *Seed Biology*, ed. T. T. Kozlowski, pp. 145–245. New York: Academic Press.

Harrington, J. F. (1973). Problems of seed storage. In *Seed Ecology*, ed. W. Heydecker, University Park, Pennsylvania: Pennsylvania State University Press.

Harris, J., H. Bentham & P. Birch. (1991). Soil microbial community provides index to progress, direction and restoration. *Restoration and Management Notes*, 9, 133–135.

Harris, L. D. (1984). *The Fragmented Forest: Island Biogeography Theory and the Preservation of Biotic Diversity*, Chicage: University of Chicago Press.

Harrison, P. (1992). *The Third Revolution: Population, Environment and a Sustainable World*, Middlesex, England: Penguin Books.

Hart, P. B. S., J. A. August & A. W. West. (1989). Long-term consequences of topsoil mining on biological and physical characteristics of two New Zealand loessial soils under grazed pasture. *Land Degradation and Rehabilitation*, 1, 77–88.

Hartmann, H. T., D. E. Kester, F. T. Davies, Jr. & R. L. Geneve. (1997). *Plant Propagation: Principles and Practices*, Upper Saddle River, New Jersey: Prentice Hall.

Haselwandter, K. & G. D. Bowen. (1996). Mycorrhizal relations in trees for agroforestry and land rehabilitation. *Forest Ecology and Management*, 81, 1–18.

Heady, H. F. (1975). *Rangeland Management*, New York: McGraw-Hill.

Heede, B. H. (1976). *Gully development and control: the status of our knowledge*. Paper RM-169. Fort Collins, Colorado: U.S. Department of Agriculture, Forest Service, Rocky Mountain Forest and Range Experiment Station.

Heichel, G. H. (1985). Symbiosis: nodule bacteria and leguminous plants. In *Forages: The Science of Grassland Agriculture*, ed. M. E. Barnes, R. F. Barnes & D. S. Metcalfe, pp. 64–71. Ames, Iowa: Iowa State University Press.

Herbel, C. H., G. H. Abernathy, C. C. Yarbrough & D. K. Gardner. (1973). Rootplowing and seeding arid rangeland in the southwest. *Journal of Range Management*, **26**, 193–197.

Herrera, M. A., C. P. Salamanca & J. M. Barea. (1993). Inoculation of woody legumes with selected arbuscular mycorrhizal fungi and rhizobia to recover desertified Mediterranean ecosystems. *Applied Environmental Microbiology*, **59**, 129–133.

Herrick, J. E., K. M. Havstad & D. P. Coffin. (1997). Rethinking remediation technologies for desertified landscapes. *Journal of Soil and Water Conservation*, **52**, 220–225.

Heske, E. J., J. H. Brown & Q. Guo. (1993). Effects of kangaroo rat exclusion on vegetation structure and plant species diversity in the Chihuahuan Desert. *Oecologia*, **95**, 520–524.

Hirose, T. & M. Tateno. (1984). Soil nitrogen patterns induced by colonization of *Polygonum cuspidatum* on Mt. Fuji. *Oecologia*, **61**, 218–223.

Hobbie, S. E. (1992). Effects of plant species on nutrient cycling. *TREE (Trends in Ecology and Evolution)*, **7**, 336–339.

Hobbs, R. A. & H. A. Mooney. (1993). Restoration ecology and invasions. In *Nature Conservation 3: Reconstruction of Fragmented Ecosystems, Global and Regional Perspectives*, ed. D. A. Saunders, R. J. Hobbs & P. R. Erlich, pp. 127–133. Chipping Norton, New South Wales, Australia: Surrey Beatty and Sons.

Hobbs, R. J. (1992a). Function of biodiversity in Mediterranean ecosystems in Australia: definitions and background. In *Biodiversity of Mediterranean Ecosystems in Australia*, ed. R. J. Hobbs, pp. 1–25. Chipping Norton, New South Wales: Surrey Beatty and Sons.

Hobbs, R. J. (1992b). Is biodiversity important for ecosystem functioning? Implications for research and management In *Biodiversity of Mediterranean Ecosystems in Australia*, ed. R. J. Hobbs, pp. 211–245. Chipping Norton, New South Wales: Surrey Beatty and Sons.

Hobbs, R. J. (1993). Effects of landscape fragmentation on ecosystem processes in the western Australian wheatbelt. *Biological Conservation*, **64**, 193–201.

Hobbs, R. J. & L. Atkins. (1988). Effect of disturbance and nutrient addition on native and introduced annuals in plant communities in the western Australian wheatbelt. *Ecology*, **13**, 171–179.

Hobbs, R. J. & D. A. Norton. (1996). Towards a conceptual framework for restoration ecology. *Restoration Ecology*, **4**, 93–110.

Hobbs, R. J. & D. A. Saunders (eds) (1993). *Reintegrating Fragmented Landscapes*, New York: Springer-Verlag.

Hobbs, R. J., D. A. Saunders & G. W. Arnold. (1993). Integrated landscape ecology: a Western Australian perspective. *Biological Conservation*, **64**, 231–238.

Hobbs, R. J., D. A. Saunders & B. M. T. Hussey. (1990). Nature conservation: the role of corridors. *AMBIO*, **19**, 94–95.

Hodkinson, D. J. & K. Thompson. (1997). Plant dispersal – the role of man. *Journal of Applied Ecology*, **34**, 1484–1496.

Hoitinek, H. A. J., M. E. Watson & P. Sutton. (1982). Reclamation of abandoned mine land with papermill sludge. *Abandoned Mine Reclamation Symposium*, ed. N. E. Smeck & P. Sutton, pp. 5–1 to 5–6. Columbus, Ohio: Ohio State University.

Holden, M. & C. Miller. (1993). New arid land revegetation techniques at Joshua Tree National Monument. *Wildland Shrub and Arid Land Restoration Symposium*, INT-GTR-315, ed. B. Roundy, E. D. McArthur, J. S. Haley & D. K. Mann, pp. 99–101. Las Vegas, Nevada: U.S. Department of Agriculture, Forest Service, Intermountain Research Station.

Holechek, J. L., R. D. Pieper & C. H. Herbel. (1989). *Range Management: Principles and Practices*, Englewood Cliffs, New Jersey: Prentice-Hall, Inc.

Hollick, M. (1993). Self-organizing systems and environmental management. *Environmental Management*, **17**, 621–628.

Holling, C. S. (1992). Cross-scale morphology, geometry, and dynamics of ecosystems. *Ecological Monographs*, **62**, 447–502.

Holmgren, M., M. Scheffer & M. A. Huston. (1997). The interplay of facilitation and competition in plant communities. *Ecology*, **78**, 1966–1975.

Hoogmoed, W. B. & L. Stroosnijder. (1984). Crust formation on sandy soils in the Sahel. I. Rainfall and infiltration. *Soil and Tillage Research*, **4**, 5–24.

Hooper, D. U. & P. M. Vitousek. (1998). Effects of plant composition and diversity on nutrient cycling. *Ecological Monographs*, **68**, 121–149.

Howell, D. (1976). Observations on the role of grazing animals in revegetating problem patches of veld. *Proceedings of the Grassland Society of South Africa*, **11**, 59–63.

Hudson, N. (1995). *Soil Conservation*, Ames Iowa: Iowa State University Press.

Huenneke, L. F., S. P. Hamburg, R. Koide, H. A. Mooney & P. M. Vitousek. (1990). Effects of soil resources on plant invasion and community structure in Californian serpentine grassland. *Ecology*, **71**, 478–491.

Hull, A. C., Jr. (1959). Pellet seeding of wheatgrass on southern Idaho rangelands. *Journal of Range Management*, **12**, 155–163.

Hull, A. C., Jr. (1970). Grass seedling emergence and survival from furrows. *Journal of Range Management*, **23**, 421–424.

Hull, A. C., Jr. (1972). Seeding rates and spacings for rangelands in southeastern Idaho and northern Utah. *Journal of Range Management*, **25**, 50–53.

Hull, A. C., Jr., R. C. Holmgren, W. H. Berry & J. A. Wagner. (1963). *Pellet Seeding on Western Rangelands*. Miscellaneous publication 922. Washington, D.C.: U.S. Department of Agriculture, Agriculture Research Service and Forest Service in cooperation with U.S. Department of Interior, Bureau of Land Management and Bureau of Indian Affairs.

Hull, A. C., Jr. & G. H. Klomp. (1966). Longevity of crested wheatgrass in the sagebrush-grass type in southern Idaho. *Journal of Range Management*, 19, 257–262.

Hull, A. C., Jr. & G. J. Klomp. (1967). Thickening and spread of crested wheatgrass stands on southern Idaho ranges. *Journal of Range Management*, 20, 222–227.

Humphreys, G. S. (1981). The rate of ant mounding and earthworm casting near Syndey, New South Wales. *Search*, 12, 129–131.

Huston, M. A. (1997). Hidden treatments in ecological experiments – re-evaluating the ecosystem function of biodiversity. *Oecologia*, 110, 449–460.

Hyder, D. N., D. E. Booster, F. A. Sneva, W. A. Sawyer & J. B. Rodgers. (1961). Wheeltrack planting on sagebrush-bunchgrass range. *Journal of Range Management*, 14, 220–224.

Ingram, R. E. & J. K. Detling. (1984). Plant-herbivore interactions in a North American mixed-grass prairie. III. Soil nematode populations and root biomass on *Cynomys ludovicianus* populations and adjacent uncolonized areas. *Oecologia*, 63, 307–313.

Insam, H. & K. H. Domsch. (1988). Relationship between soil organic carbon and microbial biomass on chronosequences of reclamation sites. *Microbial Ecology*, 15, 177.

Insam, H. & K. Haselwandter. (1989). Metabolic quotient of the soil microflora in relationship to plant succession. *Oecologia*, 79, 147–178.

Isichei, A. O. (1990). The role of algae and cyanobacteria in arid lands. A review. *Arid Soil Research and Rehabilitation*, 4, 1–17.

IUCN. (1983). *World Conservation Strategy*, Gland, Switzerland: United Nations, International Union for the Conservation of Nature and Natural Systems.

Jackson, D. R., W. J. Selvidge & B. S. Ausmus. (1978). Behaviour of heavy metals in forced microcosms. I. Effects on nutrient cycling processes. *Water Air and Soil Pollution*, 11, 13–18.

Jackson, L. L. (1992). The role of ecological restoration in conservation biology. In *Conservation Biology*, ed. P. L. Fiedler & S. K. Jain, pp. 433–451. New York: Chapman and Hall.

Jacobson, T. L. C. & B. L. Welch. (1987). Planting depth of 'Hobble Creek' mountain big sagebrush seed. *Great Basin Naturalist*, 47, 497–499.

Janos, D. P. (1980). Mycorrhizae influence tropical succession. *Biotropica*, 12, 56–64.

Janzen, D. H. (1988a). Guanacaste national park: tropical ecological and biocultural restoration. In *Rehabilitating Damaged Ecosystems*, ed. J. Cairns,
Jr., pp. 143–192. Boca Raton, Florida: CRC Press.

Janzen, D. H. (1988b). Tropical ecological and biocultural restoration. *Science*,
239, 243–244.

Jarrell, W. M. & R. A. Virginia. (1990). Soil cation accumulation in a mesquite
woodland: sustained production and long-term estimates of water use and
nitrogen fixation. *Journal of Arid Environments*, **18**, 51–56.

Jastrow, J. D. (1987). Changes in soil aggregation associated with tallgrass
prairie restoration. *American Journal of Botany*, **74**, 1656–1664.

Jeffries, R. A., A. D. Bradshaw & P. D. Putwain. (1981). Growth, nitrogen accumulation and nitrogen transfer by legume species established on mine
spoils. *Journal of Applied Ecology*, **18**, 945–956.

Jenkins, F. D. & A. Ayanaba. (1979). Decomposition of carbon-14 labelled
plant material under tropical conditions. *Soil Science Society of America
Journal*, **41**, 912–916.

Jenkins, M. B., R. A. Virginia & W. M. Jarrell. (1987). Rhizobial ecology of the
woody legume mesquite (*Prosopis glandulosa*) in a Sonoran Desert arroyo.
Plant and Soil, **105**, 105–120.

Jha, A. K. & J. S. Singh. (1992). Influence of microsites on redevelopment of
vegetation on coalmine spoils in a dry tropical environment. *Journal of
Environmental Management*, **36**, 95–116.

Johnson, D. A., K. H. Asay, L. L. Tiezen, J. R. Ehleringer & P. G. Jefferson.
(1990). Carbon isotope discrimination: potential in screening cool-season
grasses for water-limited environments. *Crop Science*, **30**, 338–343.

Johnson, H. B. & H. S. Mayeux. (1992). Viewpoint: a view on species additions
and deletions and the balance of nature. *Journal of Range Management*, **45**,
322–333.

Johnson, K. H., K. A. Vogt, H. J. Clark, O. J. Schmitz & D. J. Vogt. (1996).
Biodiversity and productivity and stability of ecosystems. *TREE (Trends
in Ecology and Evolution)*, **11**, 372–377.

Johnson, R. W. & J. C. Tothill. (1985). Definition and broad geographic outline
of savanna lands. In *Ecology and Management of the World's Savannas*, ed. J.
C. Tothill & J. J. Mott, pp. 1–13. Canberra: Australian Academy of Science.

Jones, C. G., J. H. Lawton & M. Shachek. (1994). Organisms as ecosystem
engineers. *OIKOS*, **69**, 373–386.

Jones, C. G., J. H. Lawton & M. Shachek. (1997). Positive and negative effects
of organisms as physical ecosystem engineers. *Ecology*, **78**, 1946–1957.

Jones, R. M., M. Noguchi & G. A. Bunch. (1991). Levels of germinable seed
in topsoil and cattle feces in legume-grass and nitrogen-fertilized pastures
in south-east Queensland. *Australian Journal of Agricultural Research*, **42**,
953–968.

Jones, R. M. & M. SimeoNeto. (1987). Recovery of pasture seed ingested by ruminants. 3. The effects of the amount of seed in the diet and of diet quality on seed recovery from sheep. *Australian Journal of Experimental Agriculture*, **27**, 253–256.

Jones, T. A. (1997). Genetic considerations for native plant materials. *Using Seeds of Native Species on Rangelands*, General Technical Report INT – GTR–372, ed. N. L. Shaw & B. A. Roundy, pp. 22–25. Rapid City, South Dakota: U.S. Department of Agriculture, Forest Service, Intermountain Research Station.

Jordan, G. L. (1981). *Range seeding and brush management on Arizona rangelands*. No. T81121. Tucson, Arizona: University of Arizona Agricultural Experiment Station.

Jørgensen, S. E. & W. J. Mitsch. (1989). Ecological engineering principles. In *Ecological Engineering: an Introduction to Ecotechnology*, ed. W. J. Mitsch & S. E. Jørgensen, pp. 21–37. New York: John Wiley & Sons.

Kavia, Z. D. & L. N. Harsh. (1993). Proven technology of sand dune stabilization – a step to combat desertification. In *Afforestation of Arid Lands*, ed. A. P. Dwivedi & G. N. Gupta, pp. 79–86. Jodhpur, India: Scientific Publishers.

Keddy, P. A. (1992). Assembly and response rules: two goals for predictive community ecology. *Journal of Vegetation Science*, **3**, 157–164.

Keitt, T. H., D. L. Urban & B. T. Milne. (1997). Detecting critical scales in fragmented landscapes. *Conservation Ecology*, 1 (online), 4.

Kellman, M. & M. Kading. (1992). Facilitation of tree seedling establishment in a sand dune succession. *Journal of Vegetation Science*, **3**, 679–688.

Kelrick, M. I. & J. A. MacMahon. (1985). Nutritional and physical attributes of seeds of some common sagebrush-steppe plants: some implications for ecological theory and management. *Journal of Range Management*, **38**, 65–69.

Kelrick, M. I., J. A. MacMahon, R. R. Parmenter & D. V. Sisson. (1986). Native seed preferences of shrub-steppe rodents, birds and ants: the relationships of seed attributes and seed use. *Oecologia*, **68**, 327–337.

Kemper, D., S. Dabney, L. Kramer, D. Dominick & T. Keep. (1992). Hedging against erosion. *Journal of Soil and Water Conservation*, **47**, 284–288.

Kennenni, L. & E. v. d. Maarel. (1990). Population ecology of *Acacia tortilis* in the semi-arid region of Sudan. *Journal of Vegetation Science*, **1**, 419–424.

Kerley, G. I. H. (1991). Seed removal by rodents, birds and ants in the semi-arid Karoo, South Africa. *Journal of Arid Environments*, **20**, 63–69.

Kessler, J. J. & P. Laban. (1994). Planning strategies and funding modalities for land rehabilitation. *Land Degradation and Rehabilitation*, **5**, 25–32.

Kilcher, M. R. & D. H. Heinrichs. (1968). Rates of seeding Rambler alfalfa with dryland pasture grasses. *Journal of Range Management*, **21**, 248–249.

Kilsgaard, C. W., S. E. Greene & S. G. Stafford. (1987). Nutrient concentrations in litterfall from some western conifers with special reference to calcium. *Plant and Soil,* 102, 223–227.

Kira, T., H. Ogawa & K. Shinozaki. (1953). Intraspecific competition among higher plants. I. Competition-density-yield inter-relationships in regularly dispersed populations. *Journal of Institute Polytechnical Osaka City University,* 4, 1–16.

Kishk, M. A. (1986). Land degradation in the Nile valley. *Ambio,* 15, 226–230.

Klein, D. A., B. A. Frederick, M. Biondini & M. J. Trlica. (1988). Rhizosphere microorganism effects on soluble amino acids, sugars and organic acids in the root zone of *Agropyron cristatum, A. smithii, and Bouteloua gracilis. Plant and Soil,* 110, 19–25.

Klopatek, J. M. & W. D. Stock. (1994). Partitioning of nutrients in *Acanthosicyos horridus,* a keystone species in the Namib desert. *Journal of Arid Environments,* 26, 233–240.

Klugman, S. L., W. I. Stein & D. M. Schmitt. (1974). Seed biology. In *Seeds of Woody Plants in the United States,* ed. C. S. Schopmeyer, pp. 5–40. Washington, D.C.: U.S. Department of Agriculture, Forest Service.

Knapp, E. E. & K. J. Rice. (1994). Starting from seed: genetic issues in using native grasses for restoration. *Restoration and Management Notes,* 12, 40–45.

Knoop, W. T. & B. H. Walker. (1985). Interactions of woody and herbaceous vegetation in a southern African savanna. *Journal of Ecology,* 67, 565–577.

Knowles, P. & M. C. Grant. (1981). Genetic patterns associated with growth variability in Ponderosa pine. *American Journal of Botany,* 68, 942–946.

Knutsen, G. & B. Meeting. (1991). Microalgal mass culture and forced development of biological crust in arid lands. In *Semiarid Lands and Deserts: Soil Resource and Reclamation,* ed. J. Skujins, pp. 487–506. New York: Marcel Dekker, Inc.

Kollmann, J. & H.-P. Schill. (1996). Spatial patterns of dispersal, seed predation and germination during colonization of abandoned grassland by *Quercus petraea and Corylus avellana.Vegetatio,* 125, 193–205.

Kondolf, G. M. (1995). Five elements for effective evaluation of stream restoration. *Restoration Ecology,* 3, 133–136.

Kondolf, G. M. & E. R. Micheli. (1995). Evaluating stream restoration projects. *Environmental Management,* 19, 1–15.

Korte, N. & P. Kearl. (1993). Should restoration of small western watersheds be public policy in the United States? *Environmental Management,* 17, 729–734.

Kost, D. A., D. A. Boutelle, M. M. Larson, W. D. Smith & J. P. Vimmerstedt. (1997). Papermill sludge amendments, tree protection, and tree establishment on an abandoned coal minesoil. *Journal of Environmental Quality,* 26, 1409–1416.

Kotanen, P. M. (1997). Effects of gap area and shape on recolonization by grassland plants with differing reproductive strategies. *Canadian Journal of Botany*, 75, 352–361.

Kouwen, N. & R. M. Li. (1980). Biomechanics of vegetative channel linings. *Journal of the Hydraulics Division, American Society of Civil Engineering*, 106, 1085–1103.

Krebs, C. J. (1985). *Ecology. The Experimental Analysis of Distribution and Abundance*, New York: Harper and Row.

Laflen, J. M. & T. S. Colvin. (1981). Effect of crop residue on soil loss from continuous row cropping. *Transactions of the American Society of Agricultural Engineers*, 24, 605–609.

Lal, R. (1990). *Soil Erosion in the Tropics: Principles and Management*, New York: McGraw-Hill, Inc.

Lal, R. (1992). Restoring land degraded by gully erosion in the tropics. In *Soil Restoration: Advances in Soil Science*, ed. R. Lal & B. A. Stewart, pp. 123–152. New York: Springer-Verlag.

Lal, R. (1996). Deforestation and land-use effects on soil degradation and rehabilitation in Western Nigeria. 1. Soil physical and hydrological properties. *Land Degradation & Development*, 7, 19–45.

Lal, R. & D. J. Cummings. (1979). Clearing a tropical forest. I. Effects on soil and microclimate. *Field Crops Research*, 2, 91–197.

Lal, R., G. F. Hall & F. P. Miller. (1989). Soil degradation: I. Basic processes. *Land Degradation and Rehabilitation*, 1, 51–69.

Lal, R. & B. A. Stewart. (1992). *Soil Restoration: Advances in Soil Science*, New York: Springer-Verlag.

Lamb, D., J. Parrotta, R. Keenan & N. Tucker. (1997). Rejoining habitat remnants: restoring dagraded rainforest lands. In *Tropical Forest Remnants: Ecology, Management, and Conservation of Fragmented Communities*, ed. W. F. Laurance & R. O. Bierregaard, Jr., pp. 366–385. Chicago: The University of Chicago Press.

Langkamp, P. J. & M. J. Dalling. (1983). Nutrient cycling in a stand of *Acacia holosericea*. *Australian Journal of Botany*, 31, 141–149.

Launchbaugh, J. L. (1970). Seeding rate and first-year stand relationships for six native grasses. *Journal of Range Management*, 23, 414–417.

Laurance, W. F., S. G. Laurance, L. V. Ferreira, J. M. R.-d. Merona, C. Gascon & T. E. Lovejoy. (1997). Biomass collapse in Amazonian forest fragments. *Science*, 278, 1117–1118.

Laursen, S. B. & H. E. Hunter. (1986). *Windbreaks for Montana*. Bulletin 366. Bozeman, Montana: Montana Cooperative Extensive Service.

Lawrence, D. B., R. E. Schoenike, A. Quispel & G. Bond. (1967). The role of *Dryas drummondii* in vegetation development following ice regression at Glacier Bay, Alaska, with special reference to its nitrogen fixation by root nodules. *Journal of Ecology*, 55, 793–813.

Laycock, W. A. (1991). Stable states and thresholds of range conditions on North American rangelands: a viewpoint. *Journal of Range Management*, 44, 427–433.

Le Houérou, H. N. (1984). Rain use efficiency: a unifying concept in arid-land ecology. *Journal of Arid Environments*, 7, 213–247.

Le Houérou, H. N. (1992). The role of saltbushes (*Atriplex* spp.) in arid land rehabiltation in the Mediterranean Basin: a review. *Agroforestry Systems*, 18, 107–148.

Lee, K. E. & C. E. Prankhurst. (1992). Soil organisms and sustainable productivity. *Australian Journal of Soil Research*, 30, 855–892.

Lefroy, E. C., R. J. Hobbs & L. J. Atkins. (1991). *Revegetation Guide to the Central Wheatbelt*. Bulletin 4231. Perth: Department of Agriculture Western Australia.

Leopold, D. J. & M. K. Wali. (1992). The rehabilitation of forest ecosystems in the eastern United States and Canada. In *Ecosystem Rehabilitation. 2: Ecosystem Analysis and Synthesis*, ed. M. K. Wali, pp. 187–231. The Hague, The Netherlands: SPB Academic Publishers.

Lesica, P. & T. H. DeLuca. (1996). Long-term harmful effects of crested wheatgrass on Great Plains grassland ecosystems. *Journal of Soil and Water Conservation*, 51, 408–409.

Lewis, C. A., N. P. Lester, A. D. Bradshaw, J. E. Fitzgibbon, K. Fuller, L. Hakanson & C. Richards. (1996). Considerations of scale in habitat conservation and restoration. *Canadian Journal of Fisheries and Aquatic Sciences*, 53, 440–445.

Leyshon, A. J., M. R. Kilcher & J. D. McElgunn. (1981). Seeding rates and row spacings for three forage crops grown alone or in alternate grass-alfalfa rows in southwestern Saskatchewan. *Canadian Journal of Plant Science*, 61, 711–717.

Linhart, Y. B. (1993). Restoration, revegetation, and the importance of genetic and evolutionary perspectives. *Wildland Shrub and Arid Land Restoration Symposium*, INT-GTR-315, ed. B. Roundy, E. D. McArthur, J. S. Haley & D. K. Mann, pp. 271–287. Las Vegas, Nevada: U.S. Department of Agriculture, Forest Service, Intermountain Research Station.

Lister, N. M. E. (1998). A systems approach to biodiversity conservation planning. *Environmental Monitoring and Assessment*, 49, 123–155.

Lodge, D. M. (1993). Biological invasions: lessons for ecology. *TREE (Trends in Ecology and Evolution)*, 8, 133–137.

Logan, T. J. (1992). Reclamation of chemically degraded soils. In *Soil Restoration: Advances in Soil Science*, ed. R. Lal & B. A. Stewart, pp. 13–35. New York: Springer-Verlag.

Lohrey, R. E. (1974). *Site preparation improves survival and growth of direct-seeded pines*. Research Note SO-185. New Orleans, Louisiana: U.S. Department of Agriculture, Forest Service, Southern Forest Experiment Station.

Long, A. J. (1991). Proper planting improves performance. In *Forest Regeneration Manual*, ed. M. L. Duryea & P. M. Dougherty, pp. 303–320. Boston: Kluwer Academic Publishers.

Lonsdale, W. M. (1994). Inviting trouble: introduced pasture species in northern Australia. *Australian Journal of Ecology*, **19**, 345–354.

Loomis, R. S. & D. J. Connor. (1992). *Crop Ecology*, Cambridge, England: Cambridge University Press.

Loope, W. L. & G. F. Gifford. (1972). Influence of a soil micro-floral crust on select properties of soils under pinyon-juniper in southeastern Utah. *Journal of Soil and Water Conservation*, **27**, 164–167.

Loreau, M. (1994). Material cycling and the stability of ecosystems. *The American Naturalist*, **143**, 508–513.

Louda, S. M. (1982). Distribution ecology: variation in plant recruitment over a gradient in relation to insect seed predation. *Ecological Monographs*, **52**, 25–41.

Lovell, P. H. & P. J. Lovell. (1985). The importance of plant form as a determining factor in competition and habitat exploitation. In *Studies on Plant Demography: A Festschrift for John L. Harper*, ed. J. White, pp. 209–221. New York: Academic Press.

Lovett Doust, L. (1981). Population dynamics and local specialization in a clonal perennial (*Ranunculus repens*). I. The dynamics of ramets in contrasting habitats. *Journal of Ecology*, **69**, 743–755.

Lowery, R. F. & D. H. Gjerstad. (1991). Chemical and mechanical site preparation. In *Forest Regeneration Manual*, ed. M. L. Duryea & P. M. Dougherty, pp. 251–261. Boston: Kluwer Academic Publishers.

Luce, C. H. (1997). Effectiveness of road ripping in restoring infiltration capacity of forest roads. *Restoration Ecology*, **5**, 265–270.

Ludwig, J., D. Tongway, D. Freudenberger, J. Noble & K. Hodgkinson (eds) (1997). *Landscape Ecology Function and Management: Principles for Australia's Rangelands*, Collingwood, Victoria: CSIRO Publishing.

Ludwig, J. A. & J. M. Cornelius. (1987). Locating discontinuities along ecological gradients. *Ecology*, **68**, 448–450.

Ludwig, J. A. & D. J. Tongway. (1995). Spatial organisation of landscapes and its function in semi-arid woodlands, Australia. *Landscape Ecology*, **10**, 51–63.

Ludwig, J. A. & D. J. Tongway. (1996). Rehabilitation of semiarid landscapes in Australia. 2. Restoring vegetation patches. *Restoration Ecology*, **4**, 398–406.

Lugo, A. E. (1992). Tree plantations for rehabilitating damaged forest lands in the tropics. In *Ecosystem Rehabilitation. 2: Ecosystem Analysis and Synthesis*, ed. M. K. Wali, pp. 247–255. The Hague, The Netherlands: SPB Academic Publishing.

Lugo, A. E. (1997). The apparent paradox of reestablishing species richness on degraded lands with tree monocultures. *Forest Ecology & Management*, 99, 9–19.

Luken, J. O. (1990). *Directing Ecological Succession*, New York: Chapman and Hall.

Lusk, C. H. (1995). Seed size, establishment sites and species coexistence in a Chilean rain forest. *Journal of Vegetation Science*, 6, 249–256.

Mabutt, J. A. (1984). A new global assessment of the status and trends of desertification. *Environmental Conservation*, 11, 103–113.

MacDicken, K. G. & N. T. Vergara. (1990). *Agroforestry: Classification and Management*, New York: John Wiley and Sons.

Mack, R. N. (1981). Invasion of *Bromus tectorum* into western North America: an ecological chronicle. *Agro-Ecosystems*, 7, 145–165.

Mack, R. N. (1996). Predicting the identity and fate of plant invaders – emergent and emerging approaches. *Biological Conservation*, 78, 107–121.

MacMahon, J. A. (1987). Disturbed lands and ecological theory: an essay about a mutualistic association. In *Restoration Ecology*, ed. W. R. Jordan, M. E. Gilpin & J. D. Aber, pp. 221–237. Cambridge: Cambridge University Press.

Majer, J. D. (1989). Fauna studies and land reclamation technology – a review of the history and need for such studies. In *Animals in Primary Succession: the Role of Fauna in Reclaimed Lands*, ed. J. D. Majer, pp. 5–33. New York: Cambridge University Press.

Malcolm, C. V. (1991). Establishing shrubs in saline habits. *International Conference on Agricultural Management of Salt-affected Areas*, 1, ed. R. Choukr-Allah, pp. 351–361. Agadir, Morocco.

Malik, N. & J. Waddington. (1990). No-till pasture renovation after sward suppression by herbicides. *Canadian Journal of Plant Science*, 70, 261–267.

Marlette, G. M. & J. E. Anderson. (1986). Seed banks and propagule dispersal in crested wheatgrass stands. *Journal of Applied Ecology*, 23, 161–175.

Marquez, V. J. & E. B. Allen. (1996). Ineffectiveness of two annual legumes as nurse plants for establishment of *Artemisia californica* in coastal sage scrub. *Restoration Ecology*, 4, 42–50.

Marrs, R. H. (1993). Soil fertility and nature conservation in Europe: theoretical considerations and practical management solutions. *Advances in Ecological Research*, 24, 241–300.

Marrs, R. H. & M. W. Gough. (1989). Soil fertility – a potential problem for habitat restoration. In *Biological Habitat Reconstruction*, ed. G. P. Buckley, pp. 29–44. London: Belhaven Press.

Marrs, R. H., R. D. Roberts, R. A. Skeffington & A. D. Bradshaw. (1983). Nitrogen and the development of ecosystems. In *Nitrogen as an Ecological Factor*, ed. J. A. Lee, S. McNeill & I. H. Rorison, pp. 113–136. Oxford: Blackwell Science Publications.

Marshall, A. H. & R. E. L. Naylor. (1984). Reasons for poor establishment of direct reseeded grassland. *Annals of Applied Biology*, 105, 87–96.

Martin, A. R., R. S. Moomaw & K. P. Vogel. (1982). Warm-season grass establishment with atrazine. *Agronomy Journal*, 74, 916–920.

Masters, R. A. (1995). Establishment of big bluestem and sand bluestem cultivars with metolachlor and atrazine. *Agronomy Journal*, 87, 592–596.

Masters, R. A., S. J. Nissen, R. E. Gaussoin, D. D. Beran & R. N. Stougaard. (1996). Imidazolinone herbicides restoration of Great Plains grasslands. *Weed Technology*, 10, 392–403.

Matlock, W. G. & G. R. Dutt. (1986). *A primer on water harvesting and runoff farming*. Tucson, Arizona: Agricultural Engineering Department, University of Arizona.

McArthur, E. D. (1988). New plant development in range management. In *Vegetation Science Applications for Rangeland Analysis and Management*, ed. P. T. Tueller, pp. 81–112. Boston: Kluwer Academic Publishers.

McArthur, E. D. (1991). Shrub genetic diversity and development. *IVth International Rangeland Congress*, 1, ed. A. Gaston, M. Kernick & H.-N. L. Houérou, pp. 392–396. Montpellier, France: Association Francaise de Pastoralisme.

McArthur, E. D., J. Mudge, R. V. Buren, W. R. Anderson, S. C. Sanderson & D. G. Babbel. (1998). Randomly amplified polymorphic analysis (RAPD) of *Artemesia* subgenus *Tridentatae* species and hybrids. *Great Basin Naturalist*, 58, 12–27.

McChasney, C. J., J. M. Koch & D. T. Bell. (1995). Jarrah Forest restoration in Western Australia: canopy and topographic effects. *Restoration Ecology*, 3, 105–110.

McClanahan, T. R. & R. W. Wolfe. (1993). Accelerating forest succession in a fragmented landscape: the role of birds and perches. *Conservation Biology*, 7, 279–288.

McCook, L. J. (1994). Understanding ecological succession: causal models and theories. *Vegetatio*, 110, 115–147.

McDonald, P. M., G. O. Fiddler & H. R. Harrison. (1994). *Mulching to regenerate a harsh site: effect on Douglas-fir seedlings, forbs, grasses, and ferns*. Research Paper PSW-RP-222 Albany, California: U.S. Department of Agriculture, Forest Service, Pacific Southwest Research Station.

McFarland, M. L., D. N. Ueckert & S. Hartmann. (1987). Revegetation of oil well reserve pits in west Texas. *Journal of Range Management*, 40, 122–127.

McGinnis, W. J. (1987). Effects of hay and straw mulches on the establishment of seeded grasses and legumes on rangeland and a coal strip mine. *Journal of Range Management*, 40, 119–121.

McLendon, T. & E. F. Redente. (1991). Nitrogen and phosphorus effects on secondary succession dynamics on a semi-arid sagebrush site. *Ecology*, 72, 2016–2024.

Meeting, B. (1990). Soil algae. In *The Rhizosphere*, ed. J. M. Lynch, pp. 355–368. New York: Wiley Interscience.

Menges, E. S. (1991). Seed germination percentages increases with population size in fragmented prairie species. *Conservation Biology*, 5, 158–164.

Mertia, R. S. (1993). Role of management techniques for afforestation in arid regions. In *Afforestation of Arid Lands*, ed. A. P. Dwivedi & G. N. Gupta, pp. 73–77. Jodhpur: Scientific Publishers.

Meyer, J. L. (1997). Conserving ecosystem function. In *The Ecological Basis of Conservation: Heterogeneity, Ecosystems, and Biodiversity*, ed. S. T. A. Pickett, R. S. Ostfeld, M. Shachek & G. E. Likens, pp. 136–145. New York: Chapman & Hall.

Middleton, N. J. (1990). Wind erosion and dust-storm control. In *Techniques for Desert Reclamation*, ed. A. S. Goudie, pp. 87–108. New York: John Wiley & Sons.

Miles, J. & J. W. Kinnaird. (1979). The establishment and regeneration of birch, juniper and Scots pine in the Scottish Highlands. *Scottish Forestry*, 33, 102–119.

Miller, R. M. (1987). Mycorrhizae and succession. In *Restoration Ecology*, ed. W. R. Jordan, M. E. Gilpin & J. D. Aber, pp. 205–219. Cambridge, England: Cambridge University Press.

Milton, S. J., W. R. J. Dean, M. A. duPlessis & W. R. Siegfried. (1994). A conceptual model of arid rangeland degradation. *BioScience*, 44, 70–76.

Mitsch, W. J. (1992). Applications of ecotechnology to the creation and rehabilitation of temperate wetlands. In *Ecosystem Rehabilitation. 2. Ecosystem Analysis and Synthesis*, ed. M. K. Wali, pp. 309–331. The Hague, The Netherlands: SPB Academic Publishing.

Mitsch, W. J. & J. K. Cronk. (1992). Creation and restoration of wetlands: some design consideration for ecological engineering. In *Advances in Soil Science*, ed. R. Lal & B. A. Stewart, pp. 217–259. New York: Springer-Verlag.

Mitsch, W. J. & J. G. Gosselink. (1993). *Wetlands*, New York: Van Nostrand Reinhold.

Mitsch, W. J. & S. V. Jørgensen. (1989). Introduction to ecological engineering. In *Ecological Engineering: an Introduction to Ecotechnology*, ed. W. J. Mitsch & S. V. Jørgensen, pp. 3–12. New York: John Wiley and Sons.

Mitsch, W. J., B. C. Reeder & D. M. Klarer. (1989). The role of wetlands in the control of nutrients with a case study of western Lake Erie. In *Ecological Engineering: An Introduction to Ecotechnology*, ed. W. J. Mitsch & S. E. Jørgensen, pp. 129–158. New York: John Wiley & Sons.

Mittlebach, G. G. & K. L. Gross. (1984). Experimental studies of seed predation in old-fields. *Oecologia*, **65**, 7–13.

Mohammed, A. E., C. J. Stigter & H. S. Adam. (1996). On shelterbelt design for combating sand invasion. *Agriculture, Ecosystems and Environment*, **57**, 81–90.

Monsen, S. B. (1983). *Plants for revegetation of riparian sites within the Intermountain Region.* General Technical Report INT-157. Ogden, Utah: U.S. Department of Agriculture, Forest Service, Intermountain Forest and Range Experiment Station.

Monsen, S. B. (1985). *Seed harvesting.* Annual Report of the Vegetative Rehabilitation and Equipment Workshop 39. Washington, D.C.: U.S. Department of Agriculture, Forest Service.

Moody, M. E. & R. N. Mack. (1988). Controlling the spread of plant invasions: the importance of nascent foci. *Journal of Applied Ecology*, **25**, 1009–1021.

Morgan, J. P. (1994). Soil impoverishment: a little-known technique holds potential for establishing prairie. *Restoration and Management Notes*, **12**, 55–56.

Morgan, R. P. C. (1995). Wind erosion control. In *Slope Stabilization and Runoff Control: A Bioengineering Approach*, ed. R. P. C. Morgan & R. J. Rickson, pp. 191–220. New York: E & FN Spon.

Morgan, R. P. C. & R. J. Rickson. (1995a). Conclusions. In *Slope Stabilization and Runoff Control: A Bioengineering Approach*, ed. R. P. C. Morgan & R. J. Rickson, pp. 265–271. New York: E & FN Spon.

Morgan, R. P. C. & R. J. Rickson. (1995b). *Slope Stabilization and Runoff Control: A Bioengineering Approach*, New York: E & FN Spon.

Morgan, R. P. C. & R. J. Rickson. (1995c). Water erosion control. In *Slope Stabilization and Runoff Control: A Bioengineering Approach*, ed. R. P. C. Morgan & R. J. Rickson, pp. 133–190. New York: E & FN Spon.

Morgenson, G. (1991). *Vegetative propagation of popular and willow.* General Technical Report RM-211. Park City, Utah: U.S. Department of Agriculture, Forest Service, Rocky Mountain Forest and Range Experiment Station.

Morin, J., Y. Benyamini & A. Michaeli. (1981). The effect of raindrop impact on the dynamics of soil surface crusting and water movement in the profile. *Journal of Hydrology*, **52**, 321–326.

Morris, W. F. & D. M. Wood. (1989). The role of Lupine in succession on Mount St. Helens: facilitation or inhibition? *Ecology*, **70**, 697–703.

Morrison, D. (1987). Landscape restoration in response to previous disturbance. In *Landscape Heterogeneity and Disturbance*, ed. M. G. Turner, New York: Springer-Verlag.

Mott, J. B. & D. A. Zuberer. (1991). Natural recovery of microbial populations in mixed overburden surface-mined spoils of Texas. *Arid Soil Research and Rehabilitation*, 5, 21–34.

Mowforth, M. A. & J. P. Grime. (1989). Intra-population variation in nuclear DNA amount, cell size and growth rate in *Poa annua* L. *Functional Ecology*, 3, 289–295.

Mueggler, W. F. & J. P. Blaisdell. (1955). Effect of seeding rate upon establishment and yield of crested wheatgrass. *Journal of Range Management*, 8, 74–76.

Mueller, D. M., R. A. Bowman & W. J. McGinnies. (1985). Effects of tillage and manure on emergence and establishment of Russian wildrye in a saltgrass meadow. *Journal of Range Management*, 38, 497–500.

Munda, B. D. & S. E. Smith. (1993). Genetic variation and revegetation strategies for desert rangeland ecosystems. *Wildland Shrub and Arid Land Restoration Symposium*, INT-GTR-315, ed. B. Roundy, E. D. McArthur, J. S. Haley & D. K. Mann, pp. 288–291. Las Vegas, Nevada: U.S. Department of Agriculture, Forest Service, Intermountain Research Station.

Munshower, F. F. (1994). *Practical Handbook of Disturbed Land Revegetation*, Boca Raton, Florida: Lewis Publishers.

Murcia, C. (1997). Evaluation of Andean alder as a catalyst for the recovery of tropical cloud forests in Colombia. *Forest Ecology & Management*, 99, 163–170.

Murdoch, A. J. & R. H. Ellis. (1992). Longevity, viability and dormancy. In *Seeds: the Ecology of Regeneration in Plant Communities*, ed. M. Fenner, pp. 193–229. Wallingford, UK: CAB International.

Myers, N. (1996). Environmental services of diversity. *Proceedings of the National Academy of Science*, 93, 2764–2769.

Myers, R. J. K. & G. B. Robbins. (1991). Sustaining productive pastures in the tropics. 5. Maintaining productive sown grass pastures. *Tropical Grasslands*, 25, 104–110.

Nair, P. K. R. (1993). *An Introduction to Agroforestry*, London: Kluwer Academic Publishers.

Nelson, J. R., A. M. Wilson & C. J. Goebel. (1970). Factors influencing broadcast seeding in bunchgrass range. *Journal Range Management*, 23, 163–170.

Nepstad, D. C., C. Uhl & E. A. S. Serro. (1991). Recuperation of a degraded Amazonian landscape: forest recovery and agricultural restoration. *Ambio*, 20, 248–255.

Newman, E. J. (1988). Mycorrhizal links between plants: their functioning and ecological significance. *Advances in Ecological Research*, 18, 420–422.

Noy-Meir, I. (1973). Desert ecosystems: environment and producers. *Annual Review of Ecology and Systematics*, 4, 25–51.

NRC. (1974). *Rehabilitation potential of western coal lands*, Cambridge Massachusetts: U.S. National Research Council, Ballinger Publishing Company.

NRC. (1994). *Rangeland Health: New Methods to Classify, Inventory, and Monitor Rangelands*, Washington, D.C.: Committee on Rangeland Classification, U.S. National Research Council, National Academy Press.

Ocumpaugh, W. R., S. Archer & J. W. Stuth. (1996). Switchgrass recruitment from broadcast seed vs. seed fed to cattle. *Journal of Range Management*, **49**, 368–371.

Odum, E. P. (1969). The strategy of ecosystem development. *Science*, **164**, 262–270.

Odum, H. T. (1989). Ecological engineering and self-organization. In *Ecological Engineering: An Introduction to Ecotechnology*, ed. W. J. Mitsch & S. E. Jørgensen, pp. 79–101. New York: John Wiley & Sons.

Odum, H. Y. (1962). Man in the ecosystem. *Lockwood Conference on the Suburban Forest and Ecology*, 652, pp. 57–75. Storrs, Connecticut: Connecticut Agricultural Experiment Station.

Oomes, M. J. & W. T. Elberse. (1976). Germination of six grassland herbs in microsites with different water contents. *Journal of Ecology*, **64**, 743–755.

OTA. (1993). *Harmful Non-Indigenous Species in the United States*. U.S. Government Printing Office OTA-F-565. Washington, DC: U.S. Congress, Office of Technology Assessment.

Owen-Smith, N. & S. M. Cooper. (1987). Palatability of woody plants to browsing ruminants in a South African Savanna. *Ecology*, **68**, 319–331.

Oyebande, L. & J. O. Ayoade. (1986). The watershed as a unit for planning and land development. In *Land Clearing and Development in the Tropics*, ed. R. Lal, P. A. Sanchez & J. R. W. Cummings, pp. 37–52. Boston: A. A. Balkema.

Packham, J. R., E. V. J. Cohn, P. Millett & I. C. Trueman. (1995). Introduction of plants and manipulation of field layer vegetation. In *The Ecology of Woodland Creation*, ed. R. Ferris-Kaan, pp. 129–148. New York: John Wiley & Sons.

Pahl-Worstl, C. (1995). *The Dynamic Nature of Ecosystems: Chaos and Order Entwined*, New York: John Wiley & Sons.

Pakeman, R. J. & E. Hay. (1996). Heathland seedbanks under bracken *Pteridium aquilinum* (L.) Kuhn and their importance for revegetation after bracken control. *Journal of Environmental Management*, **47**, 329–339.

Palmer, J. P. (1992). *Nutrient cycling: the key to reclamation success?* General Technical Report NE-164. Radnor, Pennsylvania: U.S. Department of Agriculture, Forest Service, Northeastern Forest Experiment Station.

Palmer, J. P. & M. J. Chadwick. (1985). Factors affecting the accumulation of nitrogen in colliery spoil. *Journal of Applied Ecology*, **22**, 249–257.

Palmer, J. P. & L. R. Iverson. (1983). Factors affecting nitrogen fixation by white clover (*Trifolium repens*) on colliery spoil. *Journal of Applied Ecology*, 20, 287–301.

Palmer, J. P., P. J. Williams, M. J. Chadwick, A. L. Morgan & C. O. Elias. (1986). Investigations into nitrogen sources and supply in reclaimed colliery spoil. *Plant and Soil*, 91, 181–184.

Parker, L. W., D. W. Freckman, Y. Steinberger, L. Diggers & W. G. Whitford. (1984). Effects of simulated rainfall and litter quantities on desert soil biota: soil respiration, microflora and protozoa. *Pedobiologia*, 26, 267–274.

Parrotta, J. A., O. H. Knowles & J. M. Wunderle. (1997). Development of floristic diversity in 10-year old restoration forests on a bauxite mined site in Amazonia. *Forest Ecology & Management*, 99, 21–42.

Pashke, M. W., T. McLendon, D. A. Klein & E. F. Redente. (1996). Effects of nitrogen availability on plant and soil communities during secondary succession on a shortgrass steppe. *Bulletin of the Ecological Society of America (Supplement)*, 77, 342.

Pastorok, R. A., A. Macdonald, J. R. Sampson, P. Wilber, D. J. Yozzo & J. P. Titre. (1997). An ecological decision framework for environmental restoration projects. *Ecological Engineering*, 9, 89–107.

Pavelic, P., K. A. Narayan & P. J. Dillon. (1997). Groundwater flow modelling to assist dryland salinity management of a coastal plain of southern Australia. *Australian Journal of Soil Research*, 35, 669–686.

Pendery, B. M. & F. D. Provenza. (1987). Interplanting crested wheatgrass with shrubs and alfalfa: effects of competition and preferential clipping. *Journal of Range Management*, 40, 514–520.

Perry, D. A. & M. P. Amaranthus. (1990). The plant-soil bootstrap: microorganisms and reclamation of degraded ecosystems. In *Environmental Restoration: Science and Strategies for Restoring the Earth*, ed. J. J. Berger, pp. 94–102. Washington, D.C.: Island Press.

Perry, D. A., M. P. Amaranthus, J. G. Borchers, S. L. Borchers & R. E. Brainerd. (1989). Bootstrapping in ecosystems. *BioScience*, 39, 230–237.

Pickett, S. T. A., S. L. Collins & J. J. Armesto. (1987a). A heirarchical consideration of causes and mechanisms of succession. *Vegetatio*, 69, 109–114.

Pickett, S. T. A., S. L. Collins & J. J. Armesto. (1987b). Models, mechanisms and pathways of succession. *The Botanical Review*, 53, 335–371.

Pickett, S. T. A. & V. T. Parker. (1994). Avoiding the old pitfalls: opportunities in a new discipline. *Restoration Ecology*, 2, 75–79.

Pickett, S. T. A., V. T. Parker & P. G. Fiedler. (1992). The new paradigm in ecology: implications for conservation biology above the species level. In *Conservation Biology: the Theory and Practice of Nature Conservation, Preservation and Management*, ed. P. G. Fiedler & S. K. Jain, pp. 65–88. London: Chapman & Hall.

Pollock, M. M., R. J. Naiman, H. E. Erickson, C. A. Johnston, J. Pastor & G. Pinay. (1995). Beaver as engineers: influences on biotic and abiotic characteristics of drainage basins. In *Linking Species and Ecosystems*, ed. C. G. Jones & J. H. Lawton, pp. 117–126. New York: Chapman & Hall.

Poorter, H. & C. Remkes. (1990). Leaf area ratio and net assimilation rates of 24 wild species differing in relative growth rate. *Oecologia*, 83, 553–559.

Poorter, H., C. Remkes & H. Lambers. (1990). Carbon and nitrogen economy of 24 wild species differing in relative growth rate. *Plant Physiology*, 94, 621–627.

Potter, K. N., T. M. Zobeck & L. J. Hagan. (1990). A microrelief index to estimate soil erodibility by wind. *Transactions of the American Society of Agricultural Engineers*, 33, 151–155.

Powell, C. L. (1980). Mycorrhizal infectivity of eroded soils. *Soil Biology and Biochemistry*, 12, 247–250.

Prajapati, M. C. & L. S. Bhushan. (1993). Afforestation of ravines. In *Afforestation of Arid Lands*, ed. A. P. Dwivedi & G. N. Gupta, pp. 243–256. Jodhpur, India: Scientific Publishers.

Prasad, R. (1993). Reclamation of degraded lands through aerial seeding. In *Afforestation of Arid Lands*, ed. A. P. Dwivedi & G. N. Gupta, pp. 239–244. Jodhpur, India: Scientific Publishers.

Prat, D. (1992). Effect of inoculation with Frankia on the growth of *Alnus* in the field. *Acta Œcologica*, 13, 463–467.

Rabeni, C. F. & S. P. Sowa. (1996). Integrating biological realism into habitat restoration and conservation strategies for small streams. *Canadian Journal of Fisheries and Aquatic Sciences*, 53 (Supplement 1), 252–259.

Radosevich, S. R. & J. S. Holt. (1984). *Weed Ecology: Implications for Vegetation Management*, New York: John Wiley and Sons.

Ray, G. J. & B. J. Brown. (1995). Restoring Caribbean dry forests: evaluation of tree propagation techniques. *Restoration Ecology*, 3, 86–94.

Read, D. J., R. Francis & R. D. Finlay. (1985). Mycorrhizal mycelia and nutrient cycling in plant communities. In *Ecological Interactions in Soil*, ed. A. H. Fitter, pp. 193–217. Oxford, England: Blackwell Scientific Publications.

Reddell, P., H. G. Diem & Y. R. Dommergues. (1991). Use of actinorhizal plants in arid and semiarid environments. In *Semiarid Lands and Deserts: Soil Resource and Reclamation*, ed. J. Skujins, pp. 469–485. Logan, Utah: Marcel Dekker, Inc.

Reddy, K. R. & P. M. Gale. (1994). Wetland processes and water quality: a symposium overview. *Journal of Environmental Quality*, 23, 875–877.

Reeves, F. B., D. Wagner, T. Moorman & J. Kiel. (1979). The role of endomycorrhizae in revegetation practices in the semi-arid. I. A comparison of incidence of mycorrhizae in severely disturbed versus natural environments. *American Journal of Botany*, 66, 6–13.

Reij, C., P. Mulder & L. Bergermann. (1988). *Water harvesting for plant production.* Technical Paper 91. Washington, D.C.: The World Bank.

Reiners, W. A., A. F. Bouman, W. F. J. Parsons & M. Keller. (1994). Tropical rain forest conversion to pasture: changes in vegetation and soil properties. *Ecological Applications,* 4, 363–377.

Rhoades, C. C. (1997). Single-tree influences on soil properties in agroforestry – lessons from natural forest and savanna ecosystems. *Agroforestry Systems,* 35, 71–94.

Rice, K. J. & E. E. Knapp. (1997). Genes on the range: ecological genetics of restoration on rangelands. *Using Seeds of Native Species on Rangelands,* General Technical Report INT–GTR–372, ed. N. L. Shaw & B. A. Roundy, pp. 21. Rapid City, South Dakota: U.S. Department of Agriculture, Forest Service, Intermountain Research Station.

Rice, R. C. & R. S. Bowman. (1988). Effect of sample size on parameter estimates in solute-transport experiments. *Soil Science,* 146, 108–112.

Ries, R. E. & L. Hofmann. (1996). Perennial grass establishment in relationship to seeding dates in the Northern Great Plains. *Journal of Range Management,* 49, 504–508.

Rietkerk, M. & J. Vandekoppel. (1997). Alternate stable states and threshold effects in semi-arid grazing systems. *OIKOS,* 79, 69–76.

Roberts, D. W. (1987). A dynamical system perspective on vegetation theory. *Vegetatio,* 69, 27–33.

Roberts, E. H. (1973). Predicting the storage life of seeds. *Seed Science and Technology,* 1, 499–514.

Roberts, R. D. & A. D. Bradshaw. (1985). The development of hydraulic seeding techniques for unstable sand slopes. Field evaluation. *Journal of Applied Ecology,* 22, 979–994.

Roberts, R. D., R. H. Marrs, R. A. Skeffington & A. D. Bradshaw. (1981). Ecosystem development on naturally colonized china clay wastes. I. Vegetation changes and overall accumulation of organic matter and nutrients. *Journal of Ecology,* 69, 153–161.

Robinson, G. R. & S. N. Handel. (1983). Forest restoration on a closed landfill: rapid addition of new species by bird dispersal. *Conservation Biology,* 7, 271–278.

Rosenberg, D. B. & S. M. Freedman. (1984). Application of a model of ecological succession to conservation and land-use management. *Environmental Management,* 11, 323–329.

Rosenweig, M. L. (1987). Restoration ecology: a tool to study population interactions? In *Restoration Ecology: A Synthetic Approach to Ecological Research,* ed. W. R. I. Jordan, M. E. Gilpin & J. D. Aber, pp. 189–203. New York: Cambridge University Press.

Ross, M. A. & J. L. Harper. (1972). Occupation of biological space during seedling establishment. *Journal of Ecology*, 60, 77–88.

Roundy, B. A. (1987). Seedbed salinity and the establishment of range plants. *Seed and Seedbed Ecology of Rangeland Plants*, ed. G. W. Frasier & R. A. Evans, pp. 68–81. Springfield, Virginia: U.S. Department of Agriculture, Agriculture Research Service, National Technical Information Service.

Roundy, B. A., L. B. Abbott & M. Livingston. (1997). Surface soil water loss after summer rainfall in a semi-desert grassland. *Arid Soil Research and Rehabilitation*, 11, 49–62.

Roundy, B. A. & C. A. Call. (1988). Revegetation of arid and semiarid range-lands. In *Vegetation Science Applications for Rangeland Analysis and Management*, ed. P. T. Tueller, pp. 607–635. Boston: Kluwer Academic Publishers.

Roundy, B. A., R. N. Keys & V. K. Winkel. (1990). Soil response to cattle trampling and mechanical seedbed preparation. *Arid Soil Research and Rehabilitation*, 4, 233–242.

Roundy, B. A., V. K. Winkel, J. R. Cox, A. K. Dobrenz & H. Tewolde. (1993). Sowing depth and soil water effects on seedling emergence and root morphology of three warm-season grasses. *Agronomy Journal*, 85, 975–982.

Rubio, H. O., M. K. Wood, M. Cardenas & B. A. Buchanan. (1989). Effect of polyacrylamide on seedling emergence of three grass species. *Soil Science*, 148, 355–360.

Rubio, H. O., M. K. Wood, M. Cardenas & B. A. Buchanan. (1990). Seedling emergence and root elongation of four grass species and evaporation from bare soil as affected by polyacrylamide. *Journal of Arid Environments*, 18, 33–41.

Rubio, H. O., M. K. Wood, M. Cardenas & B. A. Buchanan. (1992). The effect of polyacrylamide on grass emergence in southcentral New Mexico. *Journal of Range Management*, 45, 296–300.

Rumbaugh, M. D., G. Semeniuk, R. Moore & J. D. Colburn. (1965). *Travois – an alfalfa for grazing*. Bulletin 525. Brookings, South Dakota: South Dakota Agricultural Experiment Station.

Ruprecht, J. K. & N. J. Schofield. (1991). Effects of partial deforestation on hydrology and salinity in high salt storage landscapes. I. Extensive block clearing. *Journal of Hydrology*, 129, 19–38.

Ruyle, G. B., B. A. Roundy & J. R. Cox. (1988). Effects of burning on germin-ability of Lehmann lovegrass. *Journal of Range Management*, 41, 404–406.

Ruzek, L. (1994). Bioindication of soil fertility and a mathematical model for restoration assessment. *Restoration Ecology*, 2, 112–119.

Ryszkowski, L. (1989). Control of energy and matter fluxes in agricultural landscapes. *Agriculture, Ecosystems and Environment*, 27, 107–118.

Ryszkowski, L. (1992). Energy and material flows across boundaries and ecological flows. In *Landscape Boundaries: Consequences for Biotic Diversity and Ecological Flows*, ed. A. J. Hansen & F. d. Castri, pp. 270–284. New York: Springer-Verlag.

Ryszkowski, L. (1995). Managing ecosystem services in agriculural landscapes. *Nature and Resources*, **31**, 27–36.

Sackett, S., S. Haase & M. G. Harrington. (1994). Restoration of southwestern ponderosa pine ecosystems with fire. *Sustainable Ecological Systems: Implementing an Ecological Approach to Land Management*, General Technical Report RM-247, ed. W. W. Covington & L. F. DeBano, pp. 115–121. Flagstaff Arizona: US Department of Agriculture, Forest Service, Rocky Mountain Forest and Range Experiment Station.

Samson, D. A. & T. Phillippi. (1992). Granivory and competition as determinants of annual plant diversity in the Chihuahuan desert. *OIKOS*, **65**, 61–80.

Sandoval, F. M. & G. A. Reichman. (1971). Some properties of solonetzic (sodic) soil in western North Dakota. *Canadian Journal of Soil Science*, **51**, 143–155.

Santos, P. F., J. Phillips & W. G. Whitford. (1981). The role of mites and nematodes in early stages of buried litter decomposition in a desert. *Ecology*, **62**, 664–669.

Santos, P. F. & W. G. Whitford. (1981). Litter decomposition in the desert. *BioScience*, **31**, 145–146.

Santruckova, H. (1992). Microbial biomass, activity and soil respiration in relation to secondary succession. *Pedobiologia*, **36**, 341–350.

Satterlund, D. R. & P. W. Adams. (1992). *Wildland Watershed Management*, New York: John Wiley & Sons, Inc.

Saunders, D. A. & R. J. Hobbs. (1991). The role of corridors in conservation: what do we know and where do we go? In *Nature Conservation 2: The Role of Corridors*, ed. D. A. Saunders & R. J. Hobbs, pp. 421–427. Chipping Norton, New South Wales: Surrey Beatty & Sons.

Saunders, D. A., R. J. Hobbs & P. R. Erlich. (1993a). Reconstruction of fragmented ecosystems: problems and possibilities. In *Nature Conservation 3: Reconstruction of Fragmented Ecosystems*, ed. D. A. Saunders, R. J. Hobbs & P. R. Erlich, pp. 305–313. Chipping Norton, New South Wales: Surrey Beatty & Sons.

Saunders, D. A., R. J. Hobbs & P. R. Erlich. (1993b). *Repairing a Damaged World: An Outline for Ecological Restoration*, Chipping Norton, New South Wales: Surrey Beatty & Sons Limited.

Savill, P. S. (1976). The effect of drainage and ploughing of surface water gleys on rooting and wind throw of sitka spruce in Northern Ireland. *Forestry*, **49**, 133–141.

Scanlon, P. F., R. E. Byers & M. B. Moss. (1987). Protection of apple trees from deer browsing by a soap. *Virginia Journal of Science*, 38, 63.

Schaeffer, D. J., E. E. Herricks & H. W. Kerster. (1988). Ecosystem health: I. Measuring ecosystem health. *Environmental Management*, 12, 445–455.

Schaller, F. W. & P. Sutton (eds) (1978). *Reclamation of Drastically Disturbed Lands*, Madison, Wisconsin: American Society of Agronomy.

Schlesinger, W. H., J. A. Raikes, A. E. Hartley & A. E. Cross. (1996). On the spatial pattern of soil nutrients in desert ecosystems. *Ecology*, 77, 364–374.

Schlesinger, W. H., J. F. Reynolds, G. L. Cunningham, L. F. Huenneke, W. M. Jarrell, R. A. Virginia & W. G. Whitford. (1990). Biological feedbacks in global desertification. *Science*, 247, 1043–1048.

Schofield, N. J. (1992). Tree planting for dryland salinity control in Australia. *Agroforestry Systems*, 20, 1–23.

Schuman, G. E., J. E. M. Taylor, F. Rauzi & G. S. Howard. (1980). Standing stubble versus crimped straw mulch for establishing grass on mined lands. *Journal of Soil and Water Conservation*, 35, 25–27.

Schwarzenbach, F. H. (1996). Revegetation of an airstrip and dirt roads in central east Greenland. *Arctic*, 49, 194–199.

Scowcroft, P. G. (1991). Role of decaying logs and other organic seedbeds in natural regeneration of Hawaiian forest species on abandoned montane pasture. *Session on Tropical Forestry for People of the Pacific, XVII Pacific Science Congress*, ed. C. E. Conrad & L. A. Newell, pp. 67–73. Honolulu, Hawaii: U.S. Department of Agriculture, Forest Service, Pacific Southwest Research Station.

SCSA. (1982). *Resource conservation glossary*. Ankeny, Iowa: Soil Conservation Society of America.

Seneviratne, G., L. H. J. V. Holm & S. A. Kulasooriya. (1998). Quality of different mulch materials and their decomposition and N release under low moisture regimes. *Biology and Fertility of Soils*, 26, 136–140.

SER. (1994). Project policies of the Society for Ecological Restoration. *Restoration Ecology*, 2, 132–133.

Shanan, L., N. H. Tadmor, M. Evenari & P. Reiniger. (1970). Runoff farming in the desert. III. Microcatchments for improvement of desert range. *Agronomy Journal*, 62, 445–449.

Shaver, G. R. & J. M. Melillo. (1984). Nutrient budgets of marsh plants: efficiency concepts and relation to availability. *Ecology*, 65, 1491–1510.

Sheldon, J. D. & A. D. Bradshaw. (1977). The development of a hydraulic seeding technique for unstable sand slopes. I. Effects of fertilizers, mulches and stabilizers. *Journal of Applied Ecology*, 14, 905–918.

Shirley, S. (1994). *Restoring the Tallgrass Prairie: An Illustrated Manual for Iowa and the Upper Midwest*, Iowa City, Iowa: University of Iowa Press.

Siddoway, F. H., W. S. Chepil & D. V. Armbrust. (1965). Effect of kind, amount, and placement of residue on wind erosion. *Transactions of the American Society of Agricultural Engineers*, 8, 327–331.

Siddoway, F. H. & R. H. Ford. (1971). Seedbed preparation and seeding methods to establish grassed waterways. *Journal of Soil and Water Conservation*, 26, 73–76.

Simao-Neto, M., R. M. Jones & D. Ratcliff. (1987). Recovery of pasture seed ingested by ruminants. I. Seed of six tropical pasture species fed to cattle, sheep and goats. *Australian Journal of Agricultural Research*, 27, 239–246.

Singh, S. B. & K. G. Prasad. (1993). Use of mulches in dry land afforestation programme. In *Afforestation of Arid Lands*, ed. A. P. Dwivedi & G. N. Gupta, pp. 181–190. Jodhpur, India: Scientific Publishers.

Skiffington, R. A. & A. D. Bradshaw. (1981). Nitrogen accumulation in kaolin wastes in Cornwall. IV. Sward quality and the development of a nitrogen cycle. *Plant and Soil*, 62, 439–451.

Slayback, R. D. & D. R. Cable. (1970). Larger pits aid reseeding of semidesert rangeland. *Journal of Range Management*, 23, 333–335.

Smith, D. M. (1986). *The Practice of Silviculture*, New York: John Wiley & Sons.

Smith, E. M., T. H. Taylor, J. H. Casada & W. C. Templeton. (1973). Experimental grassland renovator. *Agronomy Journal*, 65, 506–508.

Smith, F. (1996). Biological diversity, ecosystem stability and economic development. *Ecological Economics*, 16, 191–203.

Smith, R. E. N., N. R. Webb & R. T. Clarke. (1991). The establishment of heathland on old fields in Dorset, England. *Biological Conservation*, 57, 221–234.

Smith, S. T. & T. C. Stoneman. (1970). *Salt movement in bare saline soils*. Technical Bulletin 4. Perth: Western Australia Department of Agriculture.

Smith, T. & M. Houston. (1989). A theory of the spatial and temporal dynamics of plant communities. *Vegetatio*, 83, 49–69.

Sneva, F. A. & L. R. Rittenhouse. (1976). *Crested wheatgrass production: impacts on fertility, row spacing, and stand age*. Technical Bulletin 135. Corvallis, Oregon: Oregon Agricultural Experiment Station.

Snow, C. S. R. & R. H. Marrs. (1997). Restoration of *Calluna* heathland on a bracken *Pteridium*-infested site in North West England. *Biological Conservation*, 81, 35–42.

Sopper, W. E. (1992). Reclamation of mine land using municipal sludge. In *Advances in Soil Science*, ed. R. Lal & B. A. Stewart, pp. 351–431. New York: Springer-Verlag.

Spitzer, H. A. (1993). Antelope Valley emergency soil erosion control. *Land and Water*, 37, 20–24.

Springfield, H.W. (1970). *Emergence and survival of winterfat seedlings from four planting depths*. Research Note TM-162. Ft. Collins, Colorado: US Department of Agriculture, Forest Service, Rocky Mountain Forest and Range Experiment Station.

Sprugel, D. G. (1991). Disturbance, equilibrium, and environmental variability: what is 'natural' vegetation in a changing environment. *Biological Conservation*, **58**, 1–18.

St. Clair, L. L., J. R. Johansen & B. L. Webb. (1986). Rapid stabilization of fire-disturbed sites using a soil crust slurry: innoculation studies. *Reclamation and Revegetation Research*, **4**, 261–269.

St. John, T. V. (1990). Mycorrhizal inoculation of container stock for restoration of self-sufficient vegetation. In *Environmental Restoration: Science and Strategies for Restoring the Earth*, ed. J. J. Berger, pp. 103–112. Washington, D. C.: Island Press.

Stanton, N. L. (1983). The effect of clipping and phytophagous nematodes on net primary production of blue grama, *Bouteloua gracilis*. *OIKOS*, **40**, 249–257.

Stanton, N. L. (1988). The underground in grasslands. *Annual Review of Ecology and Systematics*, **19**, 573–589.

Stanton, N. L., M. Allen & M. Campion. (1981). The effect of the pesticide carbofuron on soil organisms and root and shoot production in shortgrass prairie. *Journal of Applied Ecology*, **18**, 417–431.

Starchurski, A. & J. R. Zimka. (1975). Methods of studying forest ecosystems: leaf area, leaf production, and withdrawal nutrients from leaves of trees. *Ekologia Poland*, **23**, 637–648.

Steenbergh, W. F. & C. H. Lowe. (1969). Critical factors during the first years of life of the saguaro (*Cereus giganteus*) at Saguaro National Monument, Arizona. *Ecology*, **50**, 825–834.

Steffen, J. F. (1997). Seed treatment and propagation methods. In *The Tallgrass Restoration Handbook: for Prairies, Savannas, and Woodlands*, ed. S. Packard & C. F. Mutel, pp. 151–162. Washington, D. C.: Island Press.

Steinberger, Y., D. W. Freckman, L. W. Parker & W. G. Whitford. (1984). Effects of simulated rainfall and litter quantities on desert soil biota: nematodes and microarthropods. *Pedobiologia*, **26**, 267–274.

Stephenson, G. R. & A. Veigel. (1987). Recovery of compacted soil on pastures used for winter cattle feeding. *Journal of Range Management*, **40**, 46–48.

Stevens, F. R. W., D. A. Thompson & P. G. Gosling. (1990). *Research experience in direct sowing for lowland plantation establishment*. Forestry Commission Research Information Note 184. Edinburgh, Scotland: Forestry Commission.

Stevenson, M. J., J. M. Bullock & L. K. Ward. (1995). Re-creating semi-natural communities: effect of sowing rate on establishment of calcareous grassland. *Restoration Ecology*, 3, 279–289.

Stevenson, M. J., L. K. Ward & R. F. Pywell. (1997). Re-creating semi-natural communities – vacuum harvesting and hand collection of seed on calcareous grassland. *Restoration Ecology*, 5, 66–76.

Stoddard, C. H. & G. M. Stoddard. (1987). *Essentials of Forestry Practice*, New York: John Wiley & Sons.

Stuth, J. W. & B. E. Dahl. (1974). Evaluation of rangeland seedings following mechanical brush control in Texas. *Journal of Range Management*, 27, 146–149.

Stutz, H. C. (1982). Broad gene pools required for disturbed lands. *Reclamation of Mined Land in the Southwest*, ed. E. F. Aldon & W. R. Oaks, Albuquerque, New Mexico: Soil Conservation Society of America.

Stutz, H. C. & J. F. Carlson. (1985). Genetic improvement of saltbush (*Atriplex*) and other chenopods. *Range Plant Improvement Symposium. 38th Annual Meeting, Society for Range Management*, Salt Lake City, Utah: Society for Range Management.

Styczen, M. E. & R. P. C. Morgan. (1995). Engineering properties of vegetation. In *Slope Stabilization and Erosion Control: A Bioengineering Approach*, ed. R. P. C. Morgan & R. J. Rickson, pp. 5–58. New York: E & FN Spon.

Sullivan, R. P. (1979). The use of alternative foods to reduce conifer seed predation by the deer mouse (*Peromyscus maniculatus*). *Journal of Applied Ecology*, 16, 475–495.

Sullivan, T. P., L. O. Nordstrom & D. S. Sullivan. (1985). Use of predator odors as repellents to reduce feeding damage by herbivores. *Journal of Chemical Ecology*, 11, 921–935.

Sullivan, T. P. & D. S. Sullivan. (1982). The use of alternative foods to reduce lodgepole pine seed predation by small mammals. *Journal of Applied Ecology*, 19, 33–45.

Susheya, L. M. & V. I. Parfenov. (1982). The impact of drainage and reclamation on the vegetation and animal kingdoms on Byelo-Russian bogs. *Proceedings of International Scientific Workshop on Ecosystem Dynamics in Freshwater Wetlands and Shallow Water Bodies*, 1, pp. 218–226. Moscow: UNEP and SCOPE.

Swanson, F. J., T. K. Kratz, N. Caine & R. G. Woodmansee. (1988). Landform effects on ecosystem patterns and processes: geomorphic features of the earth's surface regulate the distribution of organisms and processes. *BioScience*, 38, 92–98.

Swihart, R. K. & M. R. Conover. (1990). Reducing deer damage to yews and apple trees: testing Big Game Repellent®, Ro-pel®, and soap as repellents. *Wildlife Society Bulletin*, 18, 156–162.

Szabolcs, I. (1987). The global problem of salt-affected soils. *Acta Agronomica Hungarica*, **36**, 159–172.

Szewczyk, J. & J. Szwagrzyk. (1996). Tree regeneration on rotten wood and on soil in old-growth stand. *Vegetatio*, **122**, 37–46.

Tembe, S. K. (1993). Afforestation on arid lands. In *Afforestation of Arid Lands*, ed. A. P. Dwivedi & G. N. Gupta, pp. 39–44. Jodhpur, India: Scientific Publishers.

Thomas, D. S. G. (1992). Desert dune activity: concepts and significance. *Journal of Arid Environments*, **22**, 31–38.

Thomas, D. S. G. & N. J. Middleton. (1993). Salinization: a new perspective on a major desertification issue. *Journal of Arid Environments*, **24**, 95–105.

Thompson, J. R. (1992). *Prairies, Forests and Wetlands: The Restoration of Natural Landscape Commmunities in Iowa*, Iowa City, Iowa: University of Iowa Press.

Thurow, T. L. (1991). Hydrology and erosion. In *Grazing Management: an Ecological Perspective*, ed. R. K. Heitschmidt & J. W. Stuth, pp. 141–159. Portland, Oregon: Timber Press.

Thurow, T. L., W. H. Blackburn & C. A. Taylor. (1988). Infiltration and inter-ill erosion responses to selected livestock grazing strategies, Edwards Plateau, Texas. *Journal of Range Management*, **41**, 296–302.

Thurow, T. L. & A. S. R. Juo. (1991). Integrated management of agropastoral watershed landscape: a Niger case study. *IVth International Rangeland Congress*, **2**, ed. A. Gaston, M. Kernick & H.-N. L. Houérou, pp. 765–768. Montpellier, France: Association Française de Pastoralisme.

Thurow, T. L. & A. S. R. Juo. (1995). The rationale for using a watershed as the basis for planning and development. In *Agriculture and Environment: Bridging Food Production and Environmental Protection in Developing Countries*, ed. A. S. R. Juo & R. D. Freed, pp. 93 116. Madison, Wisconsin: American Society of Agronomy, Crop Science Society of America, Soil Science Society of America.

Tiedemann, A. R. & J. O. Klemedson. (1977). Effect of mesquite trees on vegetation and soils in the desert grassland. *Journal of Range Management*, **30**, 361–367.

Tilman, D. (1984). Plant dominance along an experimental nutrient gradient. *Ecology*, **65**, 1445–1453.

Tilman, D. (1987). Secondary succession and the pattern of plant dominance along experimental nitrogen gradients. *Ecological Monographs*, **57**, 189–214.

Tilman, D. (1996). Biodiversity: population versus ecosystem stability. *Ecology*, **77**, 350–363.

Timoney, K. P. & G. Peterson. (1996). Failure of natural regeneration after clearcut logging in Wood Buffalo National Park, Canada. *Forest Ecology & Management*, **87**, 89–105.

Tisdale, J. M. & J. M. Oades. (1982). Organic matter and water stable aggregates in soils. *Journal of Soil Science*, **33**, 141–163.

Tivy, J. (1990). *Agricultural Ecology*, New York: Longman Scientific and Technical.

Toky, O. P. & R. P. Bisht. (1992). Observations on the rooting patterns of some agroforestry trees in an arid region of north-western India. *Agroforestry Systems*, **18**, 245–263.

Tongway, D. (1994). *Rangeland Soil Condition Assessment Manual*, Canberra, Australia: CSIRO Publications.

Tongway, D. (1995). *Manual for Soil Condition Assessment of Tropical Grasslands*, Canberra, Australia: CSIRO Publications.

Tongway, D. J. (1991). Functional analysis of degraded rangelands as a means of defining appropriate restoration techniques. *IVth International Rangeland Congress*, 1, ed. A. Gaston, M. Kernick & H.-N. L. Houérou, pp. 166–168. Montpellier, France: Association Française de Pastoralisme.

Tongway, D. J. & J. A. Ludwig. (1994). Small-scale resource heterogeneity in semi-arid landscapes. *Pacific Conservation Biology*, 1, 201–208.

Tongway, D. J. & J. A. Ludwig. (1996). Rehabilitation of semiarid landscapes in Australia. I. Restoring productive soil patches. *Restoration Ecology*, **4**, 388–397.

Tongway, D. J. & J. A. Ludwig. (1997a). The conservation of water and nutrients within landscapes. In *Landscape Ecology Function and Management: Principles for Australia's Rangelands*, ed. J. Ludwig, D. Tongway, D. Freudenberger, J. Noble & K. Hodgkinson, pp. 13–22. Collingwood, Victoria Australia: CSIRO Publishing.

Tongway, D. J. & J. A. Ludwig. (1997b). The nature of landscape dysfunction in rangelands. In *Landscape Ecology Function and Management: Principles for Australia's Rangelands*, ed. J. Ludwig, D. Tongway, D. Freudenberger, J. Noble & K. Hodgkinson, pp. 49–62. Collingwood, Victoria Australia: CSIRO Publishing.

Toy, T. J. & R. F. Hadley. (1987). *Geomorphology and Reclamation of Disturbed Lands*, New York: Academic Press.

Trappe, J. M. (1981). Mycorrhizae and productivity of arid and semi-arid rangelands. In *Advances in Food Producing Systems for Arid and Semi-Arid Lands*, ed. J. T. Manassah & E. J. Briskey, 753 pp. New York: Academic Press.

Tsoar, H. (1990). The ecological background, deterioration and reclamation of desert dune sand. *Agriculture, Ecosystems and Environment*, **33**, 147–170.

Turner, R. M., S. M. Alcorn, S. M. Olin & J. A. Booth. (1966). The influence of shade, soil and water on saguaro seedling establishment. *Botanical Gazette*, **127**, 95–102.

Ueckert, D. N. (1979). Impact of white grub (*Phyllophaga crinita*) on a short-grass community and evaluation of selected rehabilitation practices. *Journal of Range Management*, **32**, 445–448.

Uhl, C. (1988). Restoration of degraded lands in the Amazonian Basin. In *Biodiversity*, ed. E. O. Wilson & F. M. Peter, pp. 326–332. Washington, D.C.: National Academy Press.

UNEP. (1977). *United Nations Conference on Desertification – Desertification: an overview*, Nairobi, Kenya: United Nations Environment Program.

UNEP. (1984). *General assessment of progress in the implementation of the plan of action to combat desertification 1978–84*. Nairobi, Kenya: United Nations Environment Program.

UNEP. (1987). *Our Common Future*, New York: United Nations World Commission on Environment and Development, Oxford University Press.

Urbanska, K. M. (1995). Biodiversity assessment in ecological restoration above the timberline. *Biodiversity & Conservation*, **4**, 679–695.

Urbanska, K. M. (1997). Safe sites – interface of plant population ecology and restoration ecology. In *Restoration Ecology and Sustainable Development*, ed. K. M. Urbanska, N. R. Webb & P. J. Edwards, pp. 81–110. Cambridge: Cambridge University Press.

Ursic, K. A., N. C. Kenkel & D. W. Larson. (1997). Revegetation dynamics of cliff faces in abandoned limestone quarries. *Journal of Applied Ecology*, **24**, 289–303.

Vallentine, J. F. (1989). *Range Developments and Improvements*, New York: Academic Press.

van de Koppel, J., M. Rietkerk & F. J. Weissing. (1997). Catastrophic vegetation shifts and soil degradation in terrestrial grazing systems. *TREE (Trends in Ecology and Evolution)*, **12**, 352–356.

Van Epps, G. A. & C. M. McKell. (1978). Major criteria and procedure for selecting and establishing range shrubs as rehabilitators of disturbed lands. *First International Rangeland Congress*, **1**, pp. 352–354. Denver, Colorado.

Van Epps, G. A. & C. M. McKell. (1980). *Revegetation of disturbed sites in the salt desert range of the Intermountain West*. Land Rehabilitation Series 5. Logan, Utah: Utah Agricultural Experiment Station.

Van Lear, D. H. & T. A. Waldrop. (1991). Prescribed burning for regeneration. In *Forest Regeneration Manual*, ed. M. L. Duryea & P. M. Dougherty, pp. 235–250. Boston: Kluwer Academic Publishers.

Van Voris, P., R. V. O'Neill, W. R. Emanual & H. H. Shugart. (1980). Functional complexity and ecosystem stability. *Ecology*, **61**, 1352–1360.

Vander Wall, S. B. (1993). Cache site selection by chipmunks (*Tamias* spp.) and its influence on the effectiveness of seed dispersal in Jeffrey pine (*Pinus jeffeyii*). *Oecologia*, **96**, 246–252.

Vandermeer, J. (1989). *The Ecology of Intercropping*, New York: Cambridge University Press.

Vasek, F. C. & L. L. Lund. (1980). Soil characteristics associated with a primary succession on a Mojave Desert dry lake plant community. *Ecology*, **61**, 1013–1018.

Vetaas, O. R. (1992). Micro-site effects of trees and shrubs in dry savannas. *Journal of Vegetation Science*, **3**, 337–344.

Vinton, M. A. & I. C. Burke. (1995). Interactions between individual plant species and soil nutrient status in shortgrass steppe. *Ecology*, **76**, 1116–1133.

Virginia, R. A. (1986). Soil development under legume tree canopies. Forest Ecology and Management, **16**, 69–79.

Virginia, R. A. & W. M. Jarrell. (1983). Soil properties in a mesquite-dominated Sonoran Desert ecosystem. *Soil Science Society of America Journal*, **47**, 138–144.

Vitousek, P. M. (1990). Biological invasions and ecosystem processes: toward an integration of population biology and ecosystem studies. *OIKOS*, **57**, 7–13.

Vitousek, P. M. & H. Farrington. (1997). Nutrient limitation and soil development: experimental test of a biogeochemical theory. *Biogeochemistry*, **37**, 63–75.

Vitousek, P. M. & W. A. Reiners. (1975). Ecosystem succession and nutrient retention: a hypothesis. *BioScience*, **25**, 376–381.

Vitousek, P. M., L. R. Walker, L. D. Whiteaker, D. Mueller-Dombois & P. M. Matson. (1987). Biological invasion by *Myrica faya* alters ecosystem development in Hawaii. *Science*, **238**, 802–804.

Vogel, W. G. (1984). Planting and species selection for revegetation of abandoned acid spoils. *Conference on Reclamation of Abondoned Acid Spoils*, pp. 70–83. Osage Beach, Missouri: Missouri Department of Abandoned Acid Spoils, Land Reclamation Commission.

Von Carlowitz, P. G. & G. V. Wolf. (1991). Open-pit sunken planting: a tree establishment technique for dry environments. *Agroforestry Systems*, **15**, 17–29.

Vough, L. R. & A. M. Decker. (1983). No-till pasture renovation. *Journal of Soil and Water Conservation*, **38**, 222–223.

Waddington, J. (1992). A comparison of drills for direct seeding alfalfa into established grasslands. *Journal of Range Management*, **45**, 483–487.

Waddington, J. & K. E. Bowren. (1976). Pasture renovation by direct drilling after weed control and sward suppression by herbicides. *Canadian Journal of Plant Science*, **56**, 985–988.

Wade, G. L. (1989). Grass competition and establishment of native species from forest soil seed banks. *Landscape and Urban Planning*, **17**, 135–149.

Wali, M. K. (1992). Ecology of the rehabilitation process. In *Ecosystem Rehabilitation, 1: Policy Issues*, ed. M. K. Wali, pp. 3–23. The Hague, The Netherlands: SPB Academic Publishers.

Walker, B. H. (1992). Biological and ecological redundancy. *Conservation Biology*, **6**, 18–23.

Walker, B. H. (1993). Rangeland ecology: understanding and managing change. *Ambio*, **22**, 80–87.

Walker, L. S. & F. S. Chapin. (1986). Physiological controls over seedling growth in primary succession on an Alaskan floodplain. *Ecology*, **67**, 1508–1523.

Walker, T. W. & J. K. Syers. (1976). The fate of phosphorus during pedogenesis. *Geoderma*, **15**, 1–19.

Wallace, A. & G. A. Wallace. (1986). Effect of very low rates of synthetic soil conditioners on soils. *Soil Science*, **141**, 324–327.

Wallace, L. L. (1987). Mycorrhizae in grasslands: interactions of ungulates, fungi and drought. *New Phytologist*, **105**, 619–632.

Ward, S. C., J. M. Koch & G. L. Ainsworth. (1996). The effect of timing of rehabilitation procedures on the establishment of a Jarrah Forest after bauxite mining. *Restoration Ecology*, **4**, 19–24.

Watson, A. (1990). The control of blowing sand and mobile desert dunes. In *Techniques for Desert Reclamation*, ed. A. S. Goudie, pp. 35–86. New York: John Wiley & Sons.

Watson, M. E. & H. A. J. Hoitinek. (1985). Long-term effects of papermill sludge in stripmine reclamation. *Ohio Report*, **70**, 19–21.

Watts, J. F. & G. D. Watts. (1990). Seasonal change in aquatic vegetation and its effect on river channel form. In *Vegetation and Erosion*, ed. J. B. Thornes, pp. 257–267. Chichester: Wiley.

Weber, F. R. (1986). *Reforestation in Arid Lands*, Arlington, Virginia: Volunteers in Technical Assistance.

Weiher, E. & P. A. Keddy. (1995). The assembly of experimental wetland plant communities. *OIKOS*, **73**, 323–335.

Weiner, J. (1990). Plant population ecology in agriculture. In *Agroecology*, ed. C. R. Carrol, J. H. Vandermeer & P. M. Rosset, pp. 235–262. New York: McGraw-Hill.

Welch, T. G., B. S. Rector & J. S. Alderson. (1993). *Seeding Rangeland*. Extension Bulletin B-1379. College Station, Texas: Texas Agricultural Extension Service.

West, N. E. (1993). Biodiversity of rangelands. *Journal of Range Management*, **46**, 2–13.

West, N. E. & M. M. Caldwell. (1983). Snow as a factor in salt desert shrub vegetation patterns in Curlew Valley, Utah. *American Midland Naturalist*, **109**, 376–379.

Westman, W. A. (1990). Managing for biodiversity. *BioScience*, 40, 26–33.

Westman, W. E. (1991). Ecological restoration projects: measuring their performance. *Environmental Professional*, 13, 207–215.

Whelan, R. J. (1989). The influence of fauna on plant species composition. In *Animals in Primary Succession: The Role of Fauna in Reclaimed Lands*, ed. J. J. Majer, pp. 107–142. New York: Cambridge University Press.

Whisenant, S. G. (1990). Postfire population dynamics of *Bromus japonicus*. *American Midland Naturalist*, 123, 301–308.

Whisenant, S. G. (1993). Landscape dynamics and aridland restoration. *Wildland Shrub and Arid Land Restoration Symposium*, INT-GTR-315, ed. B. Roundy, E. D. McArthur, J. S. Haley & D. K. Mann, pp. 26–34. Las Vegas, Nevada: U.S. Department of Agriculture, Forest Service, Intermountain Research Station.

Whisenant, S. G. (1995). Initiating autogenic restoration on degraded arid lands. *Fifth International Rangeland Congress*, I, ed. N. E. West, pp. 597–598. Salt Lake City, Utah: Society for Range Management.

Whisenant, S. G. & S. H. Hartmann. (1997). Oil-field pits and pads: changing eyesores to assets. *International Petroleum Environmental Conference*, 4, ed. K. Sublette, San Antonio, Texas: University of Tulsa.

Whisenant, S. G., T. L. Thurow & S. J. Maranz. (1995). Initiating autogenic restoration on shallow semiarid sites. *Restoration Ecology*, 3, 61–67.

Whisenant, S. G. & D. Tongway. (1995). Repairing mesoscale processes during restoration. *Fifth International Rangeland Congress*, II, ed. N. E. West, pp. 62–64. Salt Lake City, Utah: Society for Range Management.

Whisenant, S. G., D. N. Ueckert & J. E. Huston. (1985). Evaluation of selected shrubs for arid and semiarid game ranges. *Journal of Wildlife Management*, 49, 524–527.

Whisenant, S. G. & F. J. Wagstaff. (1991). Successional trajectories of a grazed salt desert shrubland. *Vegetatio*, 94, 133–140.

White, J. & J. L. Harper. (1970). Correlated changes in plant size and number in plant populations. *Journal of Ecology*, 58, 467–485.

White, P. S. & J. L. Walker. (1997). Approximating natures variation – selecting and using reference information in restoration ecology. *Restoration Ecology*, 5, 338–349.

Whitford, W. G. (1978). Foraging in seed-harvester ants *Pogonomyrmex* spp. *Ecology*, 59, 185–189.

Whitford, W. G. (1988). Decomposition and nutrient cycling in disturbed arid ecosystems. In *The Reconstruction of Disturbed Arid Lands: An Ecological Approach*, ed. E. B. Allen, pp. 136–161. Boulder, Colorado: Westview Press.

Whitford, W. G. (1996). The importance of the biodiversity of soil biota in arid ecosystems. *Biodiversity and Conservation*, 5, 185–195.

Whitford, W. G., E. F. Aldon, D. W. Freckman, Y. Steinberger & L. W. Parker. (1989). Effects of organic amendments on soil biota on a degraded rangeland. *Journal of Range Management*, 42, 56–60.

Whitman, A. A., N. V. L. Brokaw & J. M. Hagan. (1997). Forest damage caused by selection logging of mahogany (*Swietenia macrophylla*) in Northern Belize. *Forest Ecology & Management*, 92, 87–96.

Whittaker, R. H. (1970). Communities and environments. In *Reclamation of Surface-Mined Lands. Volume II.*, ed. L. R. Hossner, pp. 93–129. Boca Raton, Florida: CRC Press.

Wicklow, D. T. & J. C. Zak. (1983). Viable grass seed in herbivore dung from a semi-desert grassland. *Grass and Forage Science*, 38, 25–26.

Wiedemann, H. T. & B. T. Cross. (1990). *Disk-Chain-Diker implement selection and construction*. Center Technical Report 90–1. Vernon, Texas: Texas Agricultural Experiment Station, Chillicothe-Vernon Agricultural Research and Extension Center.

Wight, J. R. & F. H. Siddoway. (1972). Improving precipitation-use efficiency on rangeland by surface modification. *Journal of Soil and Water Conservation*, 27, 170–174.

Williams, J. D., J. P. Dobrowolski & N. E. West. (1997). Microphytic crust influence on interrill erosion and infiltration capacity. *Transactions of the American Society of Agricultural Engineers*, 38, 139–146.

Wilson, E. O. & E. O. Willis. (1975). Applied biogeography. In *Ecology and Evolution of Communities*, ed. M. L. Cody & J. M. Diamond, pp. 522–534. Cambridge, Massachusetts: Harvard University Press.

Wilson, G. P. M. & D. W. Hennessy. (1977). The germination of excreted kikuyu grass seed in cattle dung pats. *Journal of Agricultural Science, Cambridge*, 88, 247–249.

Wilson, J. B., R. B. Allen & W. G. Lee. (1995a). An assembly rule in the ground and herbaceous strata of a New Zealand rain forest. *Functional Ecology*, 9, 61–64.

Wilson, J. B., R. K. Peet & M. T. Sykes. (1995b). Time and space in the community structure of a species-rich limestone grassland. *Journal of Vegetation Science*, 6, 729–740.

Wilson, J. B. & S. H. Roxburgh. (1994). A demonstration of guild-based assembly rules for a plant community, and determination of intrinsic guilds. *OIKOS*, 69, 267–276.

Wilson, S. D. & A. D. Gerry. (1995). Strategies for mixed-grass prairie restoration: herbicide, tilling, and nitrogen manipulation. *Restoration Ecology*, 3, 290–298.

Wilson, S. D. & D. Tilman. (1991). Interactive effects of fertilization and disturbance on community structure and resource availability in an old-field plant community. *Oecologia*, 88, 61–71.

Winkel,V. K.,J. C. Medrano, C. Stanley & M. D.Walo. (1993). Effects of gravel mulch on emergence of galleta grass seedlings. *Wildland Shrub and Arid Land Restoration Symposium,* INT-GTR-315, ed. B. Roundy, E. D. McArthur, J. S. Haley & D. K. Mann, pp. 130–134. Las Vegas, Nevada: U. S. Department of Agriculture, Forest Service, Intermountain Research Station.

Winkel,V. K., B. A. Roundy & J. R. Cox. (1991). Influence of seedbed micro-site characteristics on grass seedling emergence. *Journal of Range Management,* 44, 210–214.

Wischmeier,W. H. & D. D. Smith. (1978). *Predicting rainfall erosion losses – a guide to conservation planning.* Agriculture Handbook 537. Washington, D.C.: U.S. Department of Agriculture.

Wood, M. K., R. E. Eckert, Jr., W. H. Blackburn & F. F. Peterson. (1982). Influence of crusting soil surfaces on emergence and establishment of crested wheatgrass, squirreltail, thurber needlegrass and fourwing salt-bush. *Journal of Range Management,* 35, 282–287.

Woodruff, N. P. & F. H. Siddoway. (1965). A wind erosion equation. *Soil Science Society of America Proceedings,* 29, 602–608.

WRI. (1992). World Resources 1992–93, Washington, D.C.: World Resources Institute.

Wyant, J. G., R. A. Maganck & S. H. Ham. (1995). A planning and decision-making framework for ecological restoration. *Environmental Management,* 19, 789–796.

Yamada, T. & T. Kawaguchi. (1972). Dissemination of pasture plants by live-stock. II. Recovery, viability, and emergence of some pasture plant seeds passed through the digestive tract of dairy cows. *Journal of Japanese Grassland Science,* 18, 8–15.

Yates, C. J., R. J. Hobbs & R.W. Bell. (1994). Landscape-scale disturbances and regeneration in semi-arid woodlands of southwestern Australia. *Pacific Conservation Biology,* 1, 214–221.

Young, A. (1974). Some aspects of tropical soils. *Geography,* 59, 233–239.

Young, J. A., R. R. Blank,W. S. Longland & D. E. Palmquist. (1994). Seeding indian ricegrass in an arid environment in the Great Basin. *Journal of Range Management,* 47, 2–7.

Young, J. A., C. D. Clements & R. R. Blank. (1997). Influence of nitrogen on antelope bitterbrush seedling establishment. *Journal of Range Management,* 50, 536–540.

Young, J. A. & D. McKenzie. (1982). Rangeland drill. Rangelands, 4, 108–113.

Zak, J. M. & J. Wagner. (1967). Oil-base mulches and terraces as aids to tree and shrub establishment on coastal sand dunes. *Journal of Soil and Water Conservation,* 22, 198–201.

Zink, T. A., M. F. Allen, B. Heindl-Tenhunen & E. B. Allen. (1995). The effect of a disturbance corridor on an ecological reserve. *Restoration Ecology*, 3, 304–310.

Zinke, P. J. & R. L. Crocler. (1962). The influence of giant sequoa on soil properties. *Forest Science*, 8, 2–11.

Zitzer, S. F., S. R. Archer & T. W. Boutton. (1996). Spatial variability in the potential for symbiotic N_2 fixation by woody plants in a subtropical savanna ecosystem. *Journal of Applied Ecology*, 33, 1125–1136.

Index

Printed in the United States
By Bookmasters